A LIFE (UN)WORTHY OF LIVING

INTERNATIONAL LIBRARY OF ETHICS, LAW, AND THE NEW MEDICINE

VOLUME 34

The titles published in this series are listed at the end of this volume.

A Life (Un)Worthy of Living

Reproductive Genetics in Israel and Germany

by

YAEL HASHILONI-DOLEV

Academic College of Tel-Aviv-Yaffo,
Israel

Springer

A C.I.P. Catalogue record for this book is available from the Library of Congress.

ISBN 978-1-4020-5217-0 (HB)
ISBN 978-1-4020-5218-7 (e-book)

Published by Springer,
P.O. Box 17, 3300 AA Dordrecht, The Netherlands.

www.springer.com

Printed on acid-free paper

This work is dedicated to the late Prof. Wertz, a symbol of academic cooperation and generosity, as well as to my beloved spouse Ido Dolev and our two boys, Ori and Yoav, who are my pride and joy.

TABLE OF CONTENTS

ACKNOWLEDGEMENTS

It is a great pleasure to acknowledge all the people in Israel and Germany who contributed to this research throughout its different stages and helped me to accomplish the work. First among them I wish to thank Yehouda Shenhav and Haim Hazan, who supervised my Ph.D. dissertation, for their tremendous intellectual inspiration ever since I joined the department for Sociology and Anthropology at T.A. University. To Yehouda Shenhav I am grateful for an enormous contribution to this work, in all its theoretical and practical ups and downs, for his excellent critical reading and above all, for being a caring friend and mentor. Haim Hazan has always been there to support and guide me with his brilliant observations, which helped me to frame my thoughts and with his outstanding humor, wisdom and encouragement for which I wish to thank him. I am deeply thankful to Rivka Carmi, who encouraged me to study her field of medical genetics, for her continual support and open mindedness to a sociological way of thinking. Hanna Herzog has always assisted me with caring advice and trust in the importance of my work, for which I am very grateful.

I would like to thank many people in Germany, who have made this research possible, as well as making the time my family and I spent in Cologne pleasurable. First of all I would like to thank Ute Deichmann, who helped me intellectually, practically and morally and became my good friend as a wonderful by-product of this project.

Similarly, I wish to thank Benno Mueller-Hill for his support, for our interesting talks and for his courageous writing, which inspired this study. Marcus Paus, who was my assistant in Germany, made my encounter with a foreign society easy and fruitful. He was someone to consult on every possible matter and his advice was always of the best kind. Marcus Paus, Ulla Gelbert and Kornelia and Rudolph Knuebel have been like a family to me while I was staying in Germany, for which I am grateful.

Furthermore, I would like to express my deep gratitude to a network of colleague-friends who guided my work. I am thankful to Barbara Prainsack for her brilliant reading and for a very precious friendship and to Sigal Goldin, Hagai Boaz, Noga Weiner, Nurit Kirsh, Nick John, Shiri Shkedi, Carmel Shalev, Michal Frenkel, Shoham Melamed and Shay Lavi, for providing me with an intellectual support group and companionship. I am also grateful to Eva Illouz for helping me to put things together in the early stages of this work and to Larissa Remennick and Ursula Naue for their valuable comments on different parts of this work.

I wish to thank Nadav Gabai, Svetlana Bolotin, Yasmin Alkalai and Amit Kaplan for their support with the statistical calculations and Orna Donath for helping me to put some order into the bibliography. I am thankful also to Seffi Stieglitz for her endless care and support throughout my years at TA University.

Additionally, I wish to thank all the German and Israeli genetic counselors who invited me into their offices and homes and gave me so much of their time and consideration. Likewise, I wish to express my gratitude to all the counselors who took the time and effort to fill out a long questionnaire they received by mail from a complete stranger. Without them, this work would not have been possible. Zully Kohan, who headed the Israeli organization of medical geneticists, has offered me much help, for which I wish to express my appreciation.

Outside the academic world, I owe an enormous debt of gratitude to Ido Dolev, my deeply beloved spouse, for sharing my life, for following me to Germany and for supporting me throughout those years. Moreover, I wish to thank my dear parents Gad and Ilana Hashiloni, for their endless love, trust and support. Last but not least, I am grateful to the one and only Angie Silva, my children's nanny.

For funding this project, I am thankful to the Shapira Scholarship Trust awarded by the University of TA, to the Chutik scholarship, awarded by the Women's Studies Forum with the NCJW, to the DAAD (German Academic Exchange Service) short-term scholarship and to the David Horovitz Research Institute on Society and Economy Scholarship.

Yael Hashiloni-Dolev

FOREWORD

A few years ago, at the same time that I was searching for a topic for my Ph.D., I was also planning to become pregnant for the first time. As an educated upper-middle class Israeli, apart from following medical advice and taking folic acid before getting pregnant, everybody around me also went for genetic testing, either prior to pregnancy or in the first stages of pregnancy. The idea of a genetic test reminded me that 29 years ago, before having me, my parents went to a genetic counselor for advice. The reason for their worries was that a year before deciding to have another child (which turned out to be me) my eldest sister, then aged 10, was diagnosed with diabetes. Thus, my parents wanted to know the chances of their future children getting the same disease. Being told that the chances were low, they decided to have me.

And here I was 28 years later. What would I do if a genetic test existed which would tell me whether my fetus is prone to develop diabetes? Would I consider abortion? Could I theoretically abort my beloved sister, who is so many other things besides a diabetic? And what would become of me if my sister did not exist? Would I be the same person at all?

Luckily for me, no genetic test for diabetes exists and I did not have to take difficult decisions. Yet, being intimately familiar with a chronic disease made it emotionally and morally difficult for me to take genetic tests for conditions with which I was basically unfamiliar at that time. Yet, I realized my difficulties were not widely shared by others in my community, who largely accepted reproductive genetics as a scientific blessing, which might (for a while) raise the anxiety level of a pregnant woman but on the whole promises a brighter future for individual parents and families, as well as for society as a whole. Going to the library of social sciences and looking for materials about genetic counseling and Israel, I was shocked to find out how correct my intuition was and how exceptional the Israeli public and professional attitude to reproductive genetics was. The most comprehensive comparative study looking into this issue was the study of Wertz and Fletcher (Wetrz and Fletcher, 1993–95) which compared the opinions of geneticists in 37 nations around the world, concerning ethics and genetics. Since at that time Dorothy Wertz had only published a few articles based on the international survey, I had written her asking to learn some more about it.

The following data (Table 1) is extracted from the materials she sent me:

These data simply shocked and amazed me, as it demonstrated not only that Israeli geneticists were shown to be extremely enthusiastic about the personal and social uses of reproductive genetics, but also that German geneticists were found to be extremely cautious concerning the same medical technology, its uses and desirable social outcomes. In fact, Wertz and Fletcher's findings revealed that

TABLE 1. The Percentage of Geneticists Agreeing with Various Statements about the Desirable Use of Reproductive Genetics in Israel, Germany and the US

The Statement	Israel	Germany	USA
An important goal of genetic counseling is to reduce the number of deleterious genes in the population.	14%	0.2%	0.5%
Before marriage, responsible people should know whether they or their prospective partner carries a genetic disorder that could be transmitted to their children.	73%	23%	44%
People at high risk for serious disorders should not have children unless they use prenatal diagnosis and selective abortion.	38%	13%	10%
A woman should have prenatal diagnosis if medically indicated by her age and family history.	68%	34%	38%
It is not fair to a child to bring it into the world with a serious genetic disorder.	68%	18%	40%
It is not fair to a family's other children knowingly to have a child with a disability.	59%	10%	22%
It is socially irresponsible knowingly to bring an infant with a serious genetic disorder into the world in an era of prenatal diagnosis.	68%	8%	26%
The existence of people with severe disabilities makes society more rich and varied.	10%	38%	24%
N	23	255	1084

Israeli and German geneticists were placed on two extremes, one enthusiastic, the other halfhearted, concerning the uses and outcomes of reproductive genetics, while geneticists from all other "advanced liberal societies" (and hence the comparative inclusion of the US in Table 1) were placed somewhere between those two poles.

These findings made me curious to learn more about Israel and Germany. The first being my home land, the second, being the place of origin of three of my four grandparents and a country in which I had spend two influential years of my adolescence. And so, the dramatic differences between the two societies, combined with my personal background and familiarity with both of them, made me decide to make this initial curiosity the topic of my Ph.D. research. I thus became eager to answer questions such as: What is it about Israeli society that makes it endorse reproductive genetics with such enthusiasm and without criticism? What makes Israeli women say yes to prenatal genetic diagnosis and what makes the attitudes of Israeli geneticists different from their counterparts in other "advanced liberal societies"? Likewise, I wanted to understand what makes German geneticists a worldwide exception in their cautious endorsement of updated reproductive genetics, and a completely opposite picture of Israeli geneticists? What is the role of German history with its murderous racial politics (whose major victims were of course the Jews, who now make up the majority of the Israeli population), in explaining these findings and are there other factors besides the most obvious one of history, which can improve our understanding of this phenomenon?

In a second e-mail to Prof. Wertz I had asked her to learn more about the specific findings from Israel and Germany. To my great surprise and joy, she sent me all the raw data collected in these countries and encouraged me to further analyze it in my own research. I was now left with the question of where was I going to look for explanations for the reported differences and what is going to be the focus of my own work?

Seminal studies concerned with PND (prenatal diagnosis) like those of Rapp (1999) and Katz-Rothman (1986), focused on the experience of women with prenatal genetic diagnosis. As Rapp writes, these women turn into moral pioneers. Situated on a research frontier of the expanding capacity for prenatal genetic diagnosis, they are forced to judge the quality of their own fetuses and to make concrete and embodied decisions about the standards for entry into the human community (Rapp, 1999). However, for the purpose of my own study, German and Israeli women seemed to be too internally diversified as groups for me to compare. Moreover, what I really wanted to understand was not the way women justify their actions and describe their experiences, issues so brightly described by Rapp and Rothman. Rather, I wanted to primarily learn about the institutional and cultural premises and constraints that are shaping women‘s experience, most of the time without their awareness, and with what I consider to be a false claim about their supposed autonomy, since in fact, women’s options are always constrained both by the value-laden technology itself and by their society’s prevalent ways of using this technology.

Thus, I decided to put genetic counselors, who are the first contact most people have with what may be the most personal of all the advances of the genetic revolution, under the spotlight of my work. Belonging to the same international professional culture and holding the same knowledge, Israeli and German counselors have very much in common and yet, their opinions differ so sharply. Thus, the focus of my work became studying how scientific knowledge is being played out against a background of national differences and describing how local culture is shaping genetic counselors’ practice and knowledge, and *vice versa*. What I found out is that cultural ideas about fetuses, normality, health risks, fertility, science and progress, suffering, family responsibilities and the lessons to be learned from history, create very different worlds of reproductive genetics in both countries under study, which adhere to a dramatically different understanding of the cultural concept of “life” or of “a life (un)worthy of living”.

Thus, in line with contemporary thought in the sociology of science and technology, this research is an instance of how science, as a field of knowledge and practice, is culturally embedded and is not above, outside or prior to culture and for how the powerful universal explanatory claims and technological intervention of science are continually constructed and undercut by local cultures and bodies.

Moving towards the end of this foreword, I now wish to remark about the uses of language in this work.

In the tradition of the sociology of health and medicine and of the body, this work problematizes terms like "birth defects", "genetic illness", "late-term abortions", "selective abortion" (these abortions are also commonly referred to as "therapeutic", "medical" or "eugenic" abortions, as they all apply to wanted pregnancies, which are terminated due to the fetus' medical condition) and even "life" itself. Putting such terms in inverted commas serves to point out the difficulty of drawing an objective or "scientific" line between what is a unique trait or a special condition and between diseases or between "early birth" and "late abortion", which in other words, is the line between a fetus that is entitled to protection and a fetus that is not, since its future life is considered unworthy of living.

In doing this, I do not wish to totally abandon the biological reality of disease or of the different stages of pregnancy, but rather to focus attention on the fact that biological "realities" and permissible acts upon human bodies, are always socially constructed. Furthermore, by questioning such terms, I wish to respect the language and politics of the "abnormal" and the "disabled" themselves.

Another word that I do not often use in this work is eugenics. This is because this term is contaminated by its history (which is itself diversified) and hence, it nowadays often operates as a "buzz word", which serves to block a contemporary, open, moral debate about the current uses of reproductive genetics (Koch, 2004; Paul, 1992; Pritchard, 2005; Novas and Rose, 2000). As I believe that contemporary reproductive genetics has both its blessings and its burdens (Hadley, 1998) as it is both liberating and discriminating and constraining at the same time, I prefer using the term reproductive genetics and not more critical and biased terms like "voluntary eugenics"(Wertz, 1998) or "free-market eugenics". This of course does not mean those terms do not portray large parts of the picture of contemporary reproductive genetics, only that they overlook other important parts of the picture, which I do not wish to ignore. Yet, while writing about the past, the term "eugenics" is obviously more appropriate.

The discussion about the uses of language is obviously a discussion about moral attitudes and hence, it leads me to the concluding remarks of this foreword, which have to do with my personal moral judgments. Obviously, what initially made me research this topic was my uneasiness with the practices of Israeli society, with its intolerance towards the genetically deviant and its lack of critical thinking about the moral conflicts embodied in reproductive genetics. Yet, as the research advanced (and I became the mother of two boys) and after getting to know the German field of reproductive genetics, things ceased to be so simple. Whereas Israeli women are generally actively recruited (one may even say: seduced) to take genetic examinations and even to abort fetuses which are, (or are suspected to be) "genetically abnormal", their German counterparts usually simply do not know or know very little about the options opened up by this new technology. Hence,

German women today give birth to abnormal children who could have been quite easily detected in uterus but their mothers (and fathers) were never exposed to an offer to take the relevant tests. Both situations are troubling to me. Therefore, after thoroughly studying this issue, I cannot take sides with either the strong supporters of reproductive genetics or with its opponents, which all seem to me to be blind to some aspects of human suffering. Thus, I will not take a normative position on the act of prenatal genetic diagnosis itself. However, this does not mean I intend to ignore the moral realm altogether. On the contrary, I intend to engage myself with moral questions by contributing to the public debate about genetic medical technology and by pointing to the different options that post industrialized societies offer to their members, regarding the uses of reproductive genetics. This is especially important because science usually runs ahead of our social ability to make moral judgments about how far technology should be allowed to reconstruct our bodies and society (Beck, 1992), and since science and technology force us to rethink our concept of "life", a concept which is both natural and cultural. Hence, the modest purpose of this study is not to support any side or to offer guidelines for policy makers, but rather to shed light on the very difficult moral dilemmas embodied in reproductive genetics and on how two different post industrialized societies handle them.

This volume deals with the subject matter as follows

Chapter 1 investigates how "life" is managed in our modern technological era. To begin with, it looks into the relationship between technological change and social change, as it studies the institutional and cultural forces with which a medical innovation is confronted in the process of its implementation within different societies. Subsequently, the chapter discusses the interplay between global professional knowledge and culture (that of reproductive genetics and of genetic counselors) and local cultural-medical definitions. Consequently, it discusses "life" itself as a socio-cultural concept. The chapter next reviews the work of Michel Foucault, which laid the foundation for a critical discussion of the management and disciplining of "life" through modern bodily techniques. However, Foucault's thesis is somewhat problematic when applied to the study of how the concept of "life" is actually constructed in a specific social context and a distinct late-modern cultural environment. This deficiency requires the application of additional theoretical frameworks to understand the different forms that the "politics of the beginning of life" take in Israel and Germany. A Sociological discussion of the normal and the pathological, as well as of Mary Douglas' notions of the relationship between "the private body" and "the body politic", together with Rose's ideas on present-day "politics of life itself" and Giorgio Agamben's emphasis on the exclusion of life in modern societies, fill this gap and help the reader to follow the theoretical outline of the book. In its conclusion the chapter turns to explaining this book's cultural perspective and its choice of analytic tools to study the delicate and problematic concept of "life" within the two studied societies.

Chapter 2 lays out the methodological foundations of this study's comparative empirical research. These are three-fold:

A. Quantitative analysis of 295 questionnaires dealing with reproductive genetics that were responded to by genetic counselors in Germany and Israel in two time periods. The first data pool is based on materials collected by Wertz and Fletcher in 1993–95 as part of their cross-national study of ethics and genetics, which I extracted and recoded. The second database is a partial replication of the original study I conducted in 2000–01.
B. Qualitative analysis comprised 32 in-depth interviews with Israeli and German genetic counselors.
C. Textual analyses of legal, professional and newspaper materials.

Chapter 3 is dedicated to getting to know the field of reproductive genetics in both nations. It starts with an historical analysis of the evolution of Israel and Germany's reproductive genetic fields. The legislative, institutional, cultural, economic, religious and social aspects of these two societies are explored as the main argument links the political logic of reproductive genetics to the idiosyncratic history of each nation. More specifically, it discusses how actors involved in the field of reproductive genetics in Israel and Germany shape a different perspective to the disturbing question of what is a life (un)worthy of living?

Chapter 4 portrays the major findings of the genetic counselors survey in Israel and Germany. Its findings show Israeli counselors to be far more supportive of selective abortions than their German counterparts. Looking for the social factors behind these differences, the paper examines the effects of the counselors' cultural-national, institutional and personal backgrounds on their attitudes to different issues commonly understood as relevant for genetic counselors' moral practices, such as abortion, eugenics, disability, patient autonomy and the rational-scientific planning of life. It concludes by arguing that: A. Despite the nondirective ethos of genetic counseling around the world, professional practices strongly resonate with their cultural context; B. Cultural-national and religious factors affect counselors' practices over and above structural and professional differences, such as place of training and professional sub-categorizations and personal characteristics, such as gender, age and years of professional experience; C. In contrast to theories that understand science and technology to be powerful social tools, which open up options that almost deterministically become obligations later on, the findings of this survey suggest that the passing of time has not relaxed counselors' moral norms but rather made them somewhat more critical of their practice and its promises.

Chapter 5 discusses the only "remedy" for abnormalities found in fetuses that reproductive genetics can offer today; namely selective abortions. It investigates the different practices of selective and "late" selective abortions in both societies. Its major finding is that "late" selective abortions are far more common in Israel than in Germany. However, it is argued that this difference cannot be attributed solely to the legal system, since both the Israeli and the German abortion laws, despite the cultural differences that are registered in their rhetoric, open quite a wide door for "late" selective abortions. Thus, the chapter argues that "late" selective abortions are far more common in Israel than in Germany due to the professional's mediation of this process, which is conditioned by local understandings of "viability" and of maternal and fetal relationship and rights.

Chapter 6 discusses a specific group of genetic anomalies, namely sex chromosome anomalies (SCAs), whose major manifestation is future infertility. Its major finding is that the issue of whether SCAs are perceived as a highly risky medical pathology or as a normative variance is related in both cultural spaces to assumptions about the importance of fertility in human life. Whereas in Israel parenthood and the reproduction of the 'national womb' is understood to be what human life is all about, German adults do not share this assumption. Accordingly, German counselors do not support abortions on the grounds of SCAs, in contrast to their Israeli counterparts, who understand infertility to endanger the worthiness of life and thus **also** to entail a higher medical risk.

Chapter 7 examines the legal responses to the notion of "wrongful life" in Israel and Germany and shows that whereas the Israeli legal system has declared some forms of life to be "wrongful", the German legal system has not. Furthermore, it shows that this diverse understanding of the bio-cultural concept of "life" characterizes not just the German and Israeli legal systems but also the positions of genetic counselors and disability organizations in both countries. Thus, the professionals in each state seem to accept the typical logics of justification found in their respective societies concerning the moral dilemmas associated with the concept of a life (un)worthy of living, as far more Israeli than German counselors believe some forms of life to be unworthy of living. Likewise, German disability organizations reject the idea of "wrongful life" and its derivatives, namely genetic tests and selective abortions, as they tend to glorify life with disability and to fear selective abortions devalue their own lives. Alternatively, Israeli disability organizations pose no opposition to the practice of reproductive genetics, and its consequences.

Chapter 8 looks at how the concept of a life (un)worthy of living is incorporated into discussions about the "good society", family responsibilities, the relationship between generations and a moral humanistic ethos. Its findings demonstrate that whereas most Israeli counselors believe the good society should be composed of a "healthy" population, many of their German counterparts believe the opposite to be true. Thus, the conflict between individuals and society concerning this matter is formulated in a contradictory fashion. German counselors imagine a conflict between individuals' wish to avoid having abnormal children, and society's need for genetic diversity and tolerance, as opposed to Israeli counselors who imagine a conflict to result from certain parents' insistence on carrying problematic pregnancies to term, at the expense of burdening society. This difference is further complicated by the two society's understandings of the potential conflict between the future child and her family, embodied in reproductive genetics. The repeating themes in the German discourse concerning parents' choice to accept children with genetic "defects" are that those parents are moral and psychological heroes. By contrast, in Israel, parents' choice to carry problematic pregnancies to term is mainly understood by genetic counselors not only as a social wrong, but also as unfair to the future child and its siblings.

Epilogue draws upon Agamben's discussion of the ancient Greek terms that served to describe what we now call "life"; *zoē*, which expressed the simple fact of living common to all living beings (animals, men or gods) and *bios*, which indicated the form or way of living proper to an individual or a group. Accordingly, the epilogue rereads the German and Israeli politics of life by asking how these different "advanced liberal" societies incorporate *zoē* and *bios* within their bio-politics and thus how their balancing of *zoē* and *bios* affect their differential practices of reproductive genetics. Consequently, the Israeli hegemonic moral order concerning "life worthy of living" is described as "*bios* bio-politics", in which qualified life begins after birth and even then must be characterized by more than the simple fact of biological existence. On the other hand, the current German hegemonic moral order is described as "*zoē* bio-politics", in which life begins upon fertilization and in which all types of life are considered to be politically qualified. The new concepts of "*zoē* bio-politics" and "*bios* bio-politics" are suggested to serve for a less culturally-bound future study of situations in which contemporary societies have to decide about the borders between life and death and about future technological manipulations of life and their effect on what it means to be a human being who is entitled to social protection.

CHAPTER 1

THEORETICAL BACKGROUND

This study explores the social organization and implementation of reproductive genetics in two contemporary post industrialized societies, namely Germany and Israel. Prenatal genetic diagnosis is the largest current use of medical genetics. Yet, although studies of bio-ethical and social questions concerning human genetics exist in great number and variety, hardly any comparative studies have attempted to empirically examine *how* the social/cultural/historical/economic and legal conditions enable or restrict the implementation of new genetic reproductive technologies in different social settings.

In order to pursue this task, this study is guided by a host of sociological and philosophical theories, which are discussed in the following chapter. To begin with, the chapter looks into the relationship between technological change and social change, as it examines the balance between the impetus of a medical innovation and the institutional and cultural forces with which it is confronted in the process of implementation within different post industrialized societies. Following this discussion of technology and society, the chapter focuses on the interplay between global professional knowledge and culture, that of reproductive genetics and of genetic counselors and local cultural-medical definitions of "life" itself. The chapter next discusses the work of Michel Foucault, which helps to understand how modern societies discipline pregnant women and their fetuses. The subsequent section is devoted to a discussion of Foucault's approach by writers such as Nicholas Rose and Georgio Agamben. Agamben's suggestion that Foucault underestimated the extent to which sovereignty is constituted in relation to different notions of life and, based on the sovereign's decision on the value (or non value) of life as such and thus on exclusion of different forms of life, serves as a platform for questioning the differences between German and Israeli biopolitics at the beginning of life.

However, as Foucault's and Agamben's analyses are not sociological, they lack the ability to account for differences between the operations of biopolitics, i.e., of modern societies' power over life and death, within different modern states. This void is filled by a discussion of the work of Mary Douglas, who theoretically interrelates between the "private body" of women and their fetuses and the specific "body politics" of different nations. In its conclusion the chapter turns to explaining this book's cultural perspective and its choice of analytic tools for studying the delicate and problematic concept of "life" within the two studied societies.

TECHNOLOGY AND SOCIETY

Different theories have looked at the relationship between technology and society, asking two major questions: 1. How and why is technology developed? 2. What is the relationship between technological change and social change (Kline, 2001)? The founding theory in that context, the theory of technological determinism, asserts that the development of technology proceeds in an autonomous manner, determined mainly by an internal logic, with social forces accompanying it. Accordingly, technological change determines social change in a prescribed manner.[1] Other, more critical, and yet deterministic normative accounts of technology, which are often cited in relation to genetic technology, claim that society is replacing political and ethical norms with those of the technologists' and scientists' goals of efficiency and productivity. Hence, such criticism assumes that in modern societies technological discourse and practice are irresistible forces, which create a world in their own image that is global and linear and in which formal rationality prevails. It was already Weber, who, by introducing the concept of rationality, claimed that when technology and science permeate social institutions and transform them, old legitimations are destroyed (Weber, 1968). Thus, in the Western world, the progressive "rationalization" or "scientization" of society is somewhat deterministic. Correlating with these arguments, Jurgen Habermas (1999), while clearly opposing what he sees as the instrumentalization of human life by human genetic engineering, still writes that as a sociological prediction it is probably right to assume that technical reason cannot be stopped. Many years earlier Habermas wrote: "Technocratic consciousness reflects not the sundering of an ethical situation but the repression of 'ethics' as such, as a category of life… the reified models of the sciences migrate into the socio-cultural life-world and gain objective power over the latter's self-understanding" (Habermas, 1968, pp. 112–113) Likewise, Timmermans and Berg (2003) argue that genetics is often primed with a coat of dire technological deterministic colors. During the 70s, and even more so during the 80s and 90s, technological determinism has been challenged by social constructivist approaches, as numerous historians and sociologists of science and technology have published case studies in which they argued that technological innovations (Bloor, 1991; Knorr-Cetina, 1982; Latour, 1987), as well as their application (Pinch and Bijker, 1987; Staudenmair, 1985), were the outcome of negotiations among several social groups such as inventors, investors, users, government agencies, the media and others, rather than the outcome of internal technical or scientific logic. Thus, the constructivist school perceives bodies of knowledge, as well as their application, to be shaped in a complex social process. It does not regard the rational aspects of science as a self-moving and self-explanatory teleological project, as society and technology are not seen as separate objects of inquiry.

---

[1] There are also softer versions of technological determinism. A good discussion of this subject could be found in Smith, Merritt Roe and Marx, Leo. 1994. *Does Technology Drive History? The Dilemma of Technological Determinism*. Cambridge Massachusetts: The MIT Press.

In this study I wish to question the deterministic normative prophecies prevalent in the social study of genetics, by comparatively exploring how technologies of reproductive genetics are being implemented in specific cultural contexts, and by looking at how time affects genetic counselors' opinions concerning the uses of their practice. In that sense, I will be dealing with the classical question of the relationship between technological change and social change, by studying the application of knowledge in two different societies. Accordingly, my main questions will be: does an innovation once introduced into a society take on a life of its own, which determines its uses and future developments in the manner of an inescapable necessity? Or does the same technology permit alternative uses? And if so, how are the consequences of technology socially, culturally, economically, historically and politically conditioned? To be even more precise, this work's main interest is in how the symbolic-normative and historical orders effect the adoption of the same technology in two different societies, namely Israel and Germany.

Following Fischer (1992), I will claim and demonstrate that people turn new technologies to various purposes, since the application of a device does not follow straightforwardly from its instrumental logic. Thus, the inner properties of an innovation do not predetermine its application and societies experience technological development differently, according to their structure and culture. Concerning human genetics, I will demonstrate that these claims are true, especially since this field is saturated with cultural moral and ethical discourse on the meaning of "life" and about the future of our species in light of the advanced scientific innovations, a discourse which shapes the adoption of prenatal technologies in the two sites under study. However, before doing that, let me first discuss the positioning of genetic counselors, who are at the focus of this study, within a global/local context.

TECHNOLOGICAL DETERMINISM AND THE APPLICATION OF REPRODUCTIVE GENETICS IN A GLOCAL CONTEXT

For the sake of the comparative analysis, I will talk about two different layers of scientific-social contexts to which the medical genetic professionals belong. One is that of their global-professional network (which I will treat as a "constant variable"), the other is that of their local-cultural-national-scientific network (both of which are obviously not entirely homogenous). Contemporary discussion concerning globalization processes introduces the concept of glocality (Ram, 1999; Robertson, 1993, 1995), a combination of the terms "global" and "local", which describes a post-modern state of affairs in which global and local tendencies are interwoven in a dialectical and non-binary process (Gooldin, 2002). In this context, the question to be asked is: how is scientific knowledge adopted into and modified by local cultural practices and systems of meaning. As the application of the same technology in Israel and Germany differs dramatically, this study's findings show very clearly that despite the technological, medical-genetic, global common ground, the adoption of reproductive genetics in both surveyed countries is strongly influenced by the local biopolitics of the two national cultures. Furthermore, outside knowledge is always being reshaped by local meaning systems, at both the production and application levels.[2] (A further discussion of the global-local perspective can be found in chapter 4.)

Yet, having said that, I do not wish to claim that the controversies and differences discussed in this work do not share the common ground of a modern rational scientific discourse. Rather, as my findings suggest, there is not just one univocal scientific discourse or logic but a multiplicity of scientific rationalities (Toulmin, 1990). In fact, two modern societies can employ scientific reasoning differently, exactly because this kind of discourse is also social in essence and therefore cannot be separated from local-moral understandings of medico-cultural concepts, such as (in our case), the cultural meaning of "life". Thus, different societies' uses of the same scientific innovations should be understood as culturally specific. This claim becomes especially clear by using a cultural framework that does not speak in terms of society's interests (which are fundamentally rational-economic) but rather in terms of cultural systems of meaning and moral schemes. As this work demonstrates, the guiding moral projects in different societies can differ dramatically, even in two contemporary, advanced, liberal societies, which are very similar in other aspects. Hence, by comparing Israel and Germany's cultures of reproductive genetics, I wish to demonstrate that especially in the field of medical genetics, moral discussion cannot be so easily repressed. Old legitimations are not very easily destroyed. The very existence of a given technology does not determine its application and globalization does not force different post-industrialized countries onto the same path. Instead, I argue that the existence of a given technology *does not* automatically call for its application, and that technical options do not automatically become social obligations since, to stress it once again, their acceptance in different countries is preconditioned by national-cultural-moral forces, which do not easily or deterministically "surrender" to scientific reason. Rather, the interaction between the global technology and its local application, does reshape the local but not in a deterministic, isomorphic fashion.

As shown by this work, at least at this point in time, genetic technology is very differently applied in Germany and Israel. To put it briefly, in Germany today, history, religion, culture and the medical and legal systems work to restrict prenatal medical genetics, as the German culture is divided by opposition(s) to genetic techniques. On the contrary, the Israeli culture, law, medical system and Jewish religion warmly welcome prenatal medical genetics in an almost completely uncritical manner. Studying this present state of affairs does not mean that as a social scientist I can be sure that in the long run, the national differences in the uses of reproductive genetics in Israel and Germany will not pale in comparison to their similarities. They may, and they may not. In any case, this study's contribution is its careful examination of the contingencies, diversities, disjunctions, multiple oppositions and contrasting norms that tend to get pushed aside in deterministic

[2] This process works both ways, as there is constant interplay between the local and the global. Yet, this study is mainly concerned with how the local reshapes the uses made of a global technology.

narratives (Scranton, 1994), while questioning how different societies adopt new technologies. The natural/cultural concept of "life", its local meanings and its effect on the adoption of techniques of reproductive genetics is the theme of the next section.

"LIFE" AS A SOCIOLOGICAL CONCEPT

While the ability to manipulate the very material of life through contemporary technologies has destabilized the very notion of life itself (Rabinow, 1992, 1996), the dramatic rise in the medical ability to sustain the life of the genetically abnormal, combined with the even newer ability to detect these "lives" in-uterus and hence to prevent them, has created a social need for new definitions of "life". Studying the sociology of life in a certain society is thus crucial for understanding its adoption of new medical technologies. Equally, the field of reproductive genetics serves as an excellent platform for studying the sociology of "life". Yet, while the recognized experts on topics such as beginning and end of life decisions tend to come form philosophy, theology, law, biology and bioethics (Morgan, 2006), the lack of attention given to sociological and anthropological perspectives becomes more salient. Bio-ethicists think in terms of moral decisions, which are reached by autonomous moral agents, who make claims based upon logical deduction from premises. However, they are short of the socio-cultural perspective which assumes that ethical conclusions regarding life are socially produced and constructed, rather than found from contemplation. Alternatively, social scientists focus on culture and society while studying how different communities negotiate the margins of life, define "personhood" and understand how life is formed (Morgan, 2006). Thus, while studying the social aspects of human genetics in two technologically advanced societies, I will focus on the social context and constraints (professional, legal, religious, cultural), which serve as the background for decision making in this field. Consequently, the different moral positions towards prenatal genetic diagnosis and its consequences, which are studied in this work, will be understood as governed by cultural logics of justification (Boltanski and Thévenot, 1999, 2006) and social constraints and not only by the inner logic of science and technology or moral philosophy. Moreover, I will argue that the conventional ways of thinking about power and interests do not capture the rich moral and cultural meanings surrounding human genetics. This is because the moral issue as it is, always leaves a residue that cannot be attributed to other explanatory factors, such as the political or the economic. Thus, while political and economic factors will also be discussed throughout this work, we shall see that the differences between the societies in Israel and Germany concerning the application of prenatal genetic diagnosis will best be understood through the perspective of socio-cultural moral reasoning. More specifically, they will be figured out through the perspective of the value placed upon human life in all its diversity and the power involved in excluding and including certain forms and stages of "life", in the two social settings. It is therefore that cultural concepts and categories such as: "a life worth living", "wrongful life" and "viability", which are institutionally and epistemologically differently constructed in the two societies, will be the focus of the research. Consequently, I will demonstrate that the analytical categories of "life", of "a life worth living", or of "pathological" or "normal" lives are deeply rooted within a cultural context, as they are not plainly biomedical categories.

Other studies have emphasized the social embeddedness of the notion of "life" in the West, and pointed to its transformation through time and space. For example, Davison and Davis (1996), writing about cultural politics at the edge of life, explain that for cultural progressivists, life has meaning only as long as it constitutes a conscious and rational existence. Thus, people who cannot make choices are not considered fully "alive". By contrast, what they term cultural conservatives, contend that only God can give life, which reflects His image and purpose and thus, any life is always meaningful. However, these different positions are not arguing solely about the nature of "life". Rather, their ideas are competing about understandings of the nature of human community or about who is qualified for inclusion in that community and for its care and protection. "In sum", write Davison and Davis (1996), "the controversies over abortion, reproduction technologies and end-of-life issues carry many layers of meaning but at root they signify different propositions about what it means to *be* human" (Davison and Davis, 1996, p. 107). Other studies, which looked into the medical concepts of life and their transformation through time, also point to the contingency of the concept of "life". For example, Fox (1974) argued that medicine is moving from an ethic grounded in the unconditional sanctity of any life, towards an ethic based on the quality of life (see also Cohen-Almagor and Shmueli, 2000). In a similar fashion, Crane (1975) suggested that life is no longer defined biologically. Rather, the patient's capacity for social interaction is used as the new criteria for evaluating life.

My argument is that those general claims do not apply to all modern societies today and that the concepts of "worthy" or "wrongful life" or of the quality versus the sanctity or value of life carry diverse meanings in different social settings. Accordingly, the interplay between those local meanings and the common ground of a modern rational scientific discourse is responsible for the different ways in which genetic technology is applied in different cultures. Concerning the field of reproductive genetics, questions such as when does a fetus become a "life" entitled to protection, what kind of a prospective future life makes a fetus "viable", or "non viable", what are the rights of the mother in relation to her fetus and its abortion, what are the moral and legal obligations of professionals in relation to the fetus and what are the duties of society concerning such issues, become extremely important. Therefore, they are the major questions of this study, which by answering them, aims to describe the different conceptualizations of "life" in two contemporary societies. These diverse conceptualizations are thoroughly dealt with in the concluding chapter of this work, which is dedicated to a theoretical discussion of the meaning of "life" in our technological era. However, for now, suffice it to say that despite the fact that the selection among embryos in both Israel and Germany is justified according to the logic of a spectrum leading from worthy and protected life to unworthy and hence, not protected life, or even a life which should be destroyed, the placing of different lives upon this spectrum differs in Israel and Germany. Yet, before reaching any conclusions, I shall now consider repro-genetics in light of Foucauldian analysis of social power, which is centered on the way modern society, defines, manages and disciplines "life".

FOUCAULT; THE DISCIPLINING OF "LIFE" AND OF FUTURE MOTHERS

In 16th century Europe, at the time of the breakdown of the feudal system and the early development of administrative states, a new approach to social regulation and control appeared. The object of this approach, which Foucault termed "governmentality", was the population, and its practices were science and technology (Foucault, 1991). With it, the emerging European states started thinking of their citizens in terms of a "population" or "society", terms which hint at the need for intervention, monitoring, management, surveillance and protection. Exercising governmentality, or looking at the social reality through the prism of "the population", the modern governments displayed a growing concern with power over the life and welfare of people, which included a preoccupation with the fertility of populations, health and illness, patterns of diet and habitation and a general concern with people's corporeal habits and costumes (Foucault, 1981). This newly emerging power over life was termed by Foucault biopower, a power that is organized around two poles: the disciplines of the body and the regulations of the population. As aptly summarized by Sawicki (1999), the power over the individual body aims at rendering the machine-like body more powerful, productive, useful and docile. The practices of this power over the body are located within institutions such as hospitals, schools and prisons but also in the everyday activities of social agents. The other form of biopower, namely bio-politics, is a regulatory power inscribed in policies and interventions governing the population, as it involves power over birth, death and health, and over the "species body". Accordingly, it is the target of state interventions and it is what is monitored and studied in demography, public health agencies, health economics and population genetics (Sawicki, 1999). The two forms of biopower are not antithetical but rather complementary. Both are occupied with "life", with producing the individual as a normal and rational subject and citizen and with constructing deviations from normative rationality and behavioral norms as medical and mental problems. In that sense, what Foucault termed modern disciplinary technologies must always be understood as moral technologies, which are part of a modern moral project (Ophir, 2003). Consequently, such modern biopower technologies, which define "good" and "bad" "right" and "wrong", or "normal" and "pathological", recruit their subjects not by fear, force or violence but rather, by creating desires and defining the normal state against which bodies are judged and policed, both from the inside and the outside. Hence, biopower is not a repressive mechanism that works through prohibition but rather by producing new objects and subjects of knowledge, by creating desires and establishing norms. Being such a "benevolent" power, biopower is a peculiarly effective and invasive form of social control. Today's new reproductive genetic technologies are especially appropriate to be understood as part of the history of biopower regulating bodies in modern societies (Sawicki, 1999). They involve sophisticated techniques of surveillance and examination like ultrasound, blood tests and amniocentesis, that make both female, male and fetuses' bodies visible as objects and subjects of medical, as well as legal and state, control and intervention. Similarly, they form new definitions of the "normal" and the "pathological" and construct new norms of responsible and "healthy" motherhood and parenthood. Moreover, as opposed to coercive eugenic measures, which were practiced until the end of the second world war mostly in Nazi Germany, but also elsewhere like the United States and Sweden (Paul, 1998), today's reproductive genetics or what Wertz (1998) has termed "voluntary eugenics" does not build on violence. Rather, it builds on what is often understood to be the individual mothers' "autonomous" decision, or in Foucauldian terms, on her self-surveillance and self-correction to norms, in a way that combines control over the individual as well as the population's bodies. That is, since the strategies of governmentality lead individuals to police themselves, as normalized subjects who pursue their own best interests, who seek self improvement, self-satisfaction, health and happiness. Concerning mothers, nowadays, pregnant women see it as their own responsibility to govern themselves and ensure the health of their fetus by following experts' advice. In that sense they are coerced to "better" or "normalize" their children or else they will be blamed as irrational, irresponsible, or not good enough mothers.

Focusing on what responsible parenthood means in late modernity, Beck-Gernsheim (1996) points to the fact that an increasing emphasis is being placed on parent's intervention in their children's life chances. The contemporary family's size has become smaller and as a result of this tendency each child has become an object of less divided parental attention and a more scarce resource, whose success must be ensured. Moreover, children have been invested with meaning as a source of personal fulfillment and emotional authenticity for their parents. Now scarce resources and precious entities, who deserve the very best of care from their parents, children are also supposed to be bettered all the time. Hence, the quest for the "perfect child" means that children's flaws are targets for correction and for expert intervention. In that sense, parents and even more so mothers, are charged with the responsibility to adhere to norms and to "improve" their children (Beck-Gernsheim, 1996). Specifically concerning fetuses' genetics, Katz-Rothman suggests that within the framework of the practice of genetic counseling, the mothers of today are expected to see their children as products of conception, which have to be carefully examined before leaving "the factory", in what is a further commodification of the human body. Genetic counseling for that matter can be viewed as serving the function of children's quality control (Katz-Rothman, 1986). Genetic counseling and its knowledge are therefore typical cases of expert knowledge and advice that are pivotal to normalization, to governmentality and to biopower, as they provide the guidelines and advice by which populations are surveyed, compared against norms and rendered productive, and by which the late-modern subjects are fabricated within a network of instruments and techniques of power.

BIOPOLITICS AND ITS THEORETIOCAL ELABORATIONS

The by now classical Foucaudian model of biopower/biopolitics asserting that modernity is situated at the point where the simple living body becomes what is at stake in society's political strategies, has been extended and updated by different writers, which have highlighted its limitations. For example, Nikolas Rose in his famous article "The politics of life itself" (2001), contends that contemporary biopolitics can no longer be framed in terms of the "fitness of the nation". This framework is no longer relevant to "advanced liberal" societies, in which the state no longer solely administers biopolitics: it now takes place in a plural field traversed by ethics committees, professional associations, findings generated by researchers, employers and insurers, self-help organizations, religious organizations and social critiques. According to Rose, the interest of such groups as well as of the contemporary state is more in individuals or groups being "at risk", than in society at large. Additionally, contemporary bio-politics, according to Rose, is driven by economic forces, a process which leads to the breaking of traditional classifications between cure and enhancement and between commodities and that which is human. Thus, contemporary biopolitics disputes the value to be accorded to "life itself", the "quality of life", the "right to life" and so on.

A different kind of critique comes from Giorgio Agamben (1998). In contrast to Foucault, Agamben claims that the origin of the connection between politics and life is fundamental to the Western tradition and must be located much earlier than in the 16th century. He thus invokes two ancient Greek terms to describe what we now call "life"; *zoē*, or "bare life", which expressed the simple fact of living common to all living beings (animals, men or gods), and *bios*, which indicated the form or way of living proper to an individual or a group, politically and morally qualified life.[3] Today, this distinction has disappeared and can only be traced in words like *bio*graphy and *zoo*logy. Furthermore, Agamben claims that the uniqueness of modern politics is not merely its inclusion of the simple living body in the polis (as argued by Foucault), but rather the fact that the hidden foundation upon which the entire political system rests is at once the exclusion of bare life from and capturing it within the political order (Agamben, 1998). Thus, Agamben claims that the fundamental biopolitical structure of modernity is the sovereign's decision on the value (or non value) of life as such.[4] According to him, it is exactly the valorization and politicization of life characteristic of modernity, which necessarily implies a

---

[3] In the philosophical tradition it is Hannah Arendt (1998) who restored the Greek/Aristotalian terms zoe and bios in her discussion of life.

[4] Disagreeing with Foucault, Agamben does not understand social power to be diffused, as he rejects Foucault's abandonment of the Sovereign (see Ophir, 2003). As regards this work, I am overlooking the argument about the relevance of the sovereign to contemporary power relations.

locus where life ceases to be politically relevant. It thus becomes "bare life", "naked life" or the life of what he terms "homo sacer", the Greek term for a person whose life can be eliminated without punishment, or killed but not murdered or sacrificed. For that matter, the concentration camp is the biopolitical paradigmatic instance of the modern state of exception, where a suspension of rules occurs and life is captured in the political order but outside the political community. However, homo sacer or bare life do not exist only in concentration camps or in boats filled with refugees but are also the outcome of contemporary medical technology. Just like the Musselman, the over-comatose patient is an extreme embodiment of bare or naked life, the life of the homo sacer, whose biological life is separated from its *bios*, or biography (Agamben, 1998; Wynn, 2002). Agamben's philosophical project is to rethink this risky separation between *zoē* and *bios* underlying political systems in the West (Wynn, 2002). Thus, he wishes for political life to aim at not separating life from its "form of life" (Agamben, 1993), namely a life from which bare life cannot be separated, or again, a life in which *zoē* is inseparable from its *bios*. Returning to reproductive genetics, Dean (2004) writes: "Sovereignty – the power of killing – is today practiced in the biomedical domain by health professionals and administrators, by relatives and carers and by prospective parents and mothers, all under the watchful guardianship of institutional ethical committees, legal regulation and therapeutic expertise" (p. 19). Yet, while it is perfectly clear what it means to fight for life to become once again a "form of life" in refugee boats or concentration camps, when dealing with end or beginning of life decisions, I argue that what Agamben implores is no longer as obvious. Does Agamben's view lead to a radical condemnation of contemporary reproductive genetic practices? Arguing along the lines of Dean (2004) and Wynn (2002), I will suggest that the idea of mapping the zones of *bios* and *zoē* and their indistinctions may facilitate areas of analysis and contestation, rather than a dogmatic rejection of reproductive genetics. The last chapter of the book is dedicated to questioning how different post industrialized societies, such as Israel and Germany, incorporate *zoē* and *bios* within their biopolitics and thus, how their balancing of *zoē* and *bios* affect their differential practice of reproductive genetics. Having discussed theories dealing with modern politics of life itself, I contend that the approaches of Foucault, Rose and Agamben, with all their differences and similarities, tend to be blind to the fact that biopower, namely the power of modern societies over life and (following Agamben) over death, is not similar in all modern nations. Hence, they are quite limited in allowing a culturally-specific understanding of the workings of this power. Adding to this discussion, I claim that the concept of "life itself", that which is regulated by biopower, is culturally specific. Thus, I pursue a focused study of the moral technologies of the body in modern societies and their relation to local understandings of the notion of "life". My claim is that in order to specifically understand how biopower operates in different modern settings, one must look at the conceptualization of "healthy", "worthy" and "unworthy", "viable" or "non-viable" lives in different societies and that these conceptualizations explain why medical genetic innovations are differently accepted and implied in Germany and Israel. Thus, in the epilog to this book I wish to complete the concept of biopower by pointing to its exclusion of life or management of death (as does Agamben) and to its cultural designation. However, for now, my starting point will be to ask: is expert advice, and in our case genetic counseling, similar in all modern nations? To be more concrete, I will ask: do genetic counselors in Israel and Germany conceive of the "normal" fetus in a similar fashion? Or is the bio cultural concept of "life" defined alike in the two professional cultures?

MEDICAL EXPERTS' DEFINITIONS OF THE "NORMAL" AND THE "PATHOLOGICAL"

Contemporary knowledge and discourse of fetuses' "normality" and of genetic risks emerges from both expert and lay sources. But, it is experts who hold most sway because of the assumed "neutral" and "scientific" character of their knowledge.

Therefore, as opposed to studies of scholars such as Rapp (1999) and Katz-Rothman (1986), which focused on women and pointed out how the meaning of their child's normality or pathology varied, depending on the socio-cultural context and the meaning of the pregnancy in the woman's life, the focus of this work is on experts. Namely, it focuses on genetic counselors' interpretations of different genetic conditions. Thus, my first question in that regard is: where do experts' ideas about the "normality" of the personal body come from, and how "scientific" and "neutral" is their expertise? The sociological constructivist answer is that imposing a bodily-medical norm is socially and not scientifically driven and that pathology as a social object and a lived experience is not an entity but an explanatory model, which is formed by the interaction of biology and social practices and meanings (Good, 1997). It was already Canguilhem (1991), the philosopher who inspired Foucault, who argued that the definition of the "normal" and the "pathological" in medical discourse is not biomedical itself. Conversely, he claimed that an ideal of perfection hovers over the intentions of defining normality positively, since it is impossible to talk about a lack, before stating what is the human natural purpose, a statement that is moral in essence. That is, since every preference for a certain order implies a rejection of a different order. Accordingly, defining the abnormal means identifying the normative character of what is considered the "normal state" and supporting certain values over others. Thus, a medical norm, or perfect health do not simply exist in the positive sense. Their function is rather to devalue existence by stimulating its modification and correction (Canguilhem, 1991). To set a norm, to normalize, is to impose a requirement on an existence, whose variety, with regard to the requirement, presents itself as hostile, as norms are used in order to right, to square and to straighten. Thus, Caunguilhem and Foucault, who dealt with the emergence of the "normal" body in modern society and with the normative-moral demands it imposes upon individuals, have claimed that the "normal" and the "pathological" are not bio-medical concepts. Likewise disease, although there is a biological reality to it, is not "natural" and therefore above, beyond or deeper than culture. However, their work did not pay tribute to local meanings and variations imposed upon the concepts of the "normal" and the "pathological" in different cultural settings. This theoretical weakness is corrected by the discipline of sociology of health and illness, which has long been arguing that categories of disease and biomedical practices are constructed within their cultural context, as Western biomedicine (like all other forms of medicine) is grounded in local and social assumptions and as the spheres of culture and science can not be easily separated (Kleinman, 1995; Lippman, 1991; Lock, 1993; Martin, 1991; Payer, 1988; Wright and Treacher, 1982). Consequently, one cannot talk of biopolitics and its effects on fetuses, women's bodies and society, without mentioning local/national ideologies, which still justify in our times the appropriation of the private womb to the rule of society. Therefore, in this study I will show that cultural and scientific claims and expert knowledge about specific bodies' genetic "health" or "pathology" and about "worthy" and "unworthy" lives are interrelated and dependant upon national understandings of private and collective bodies. In that sense, expert knowledge is not neutral or scientific but rather culturally mediated. Just as well, it is not global or universal but rather glocal and dependant upon the interplay between global knowledge, private bodies and the body politics of different nations, which is the theme of the next section.

THE SOCIOLOGY OF BODY AND NATION

Sociologists working under the umbrella of social constructivism have long claimed that the body is socially constructed through education, work, sports and discourse[5] and that it is shaped, constrained and even invented by society (Shilling, 1993). Whereas these assumptions sometimes sound very metaphorical, in the case of the application of genetic knowledge to human bodies, they become quite literal. Contemporary science enables greater degrees of intervention into the body, as the body is increasingly becoming a phenomenon of options and choice (Giddens, 1991), as a result of developments in spheres such as biological reproduction and genetic engineering but also of plastic surgery or sports science. These novel options for molding our bodies, or our "lives", as the body is the locus where life actually takes place – have stimulated a heightened degree of reflexivity about what the body is, and an uncertainty about how it should be controlled (Shilling, 1993). This is due to the fact that the rapid development of science exceeds our ability to make moral judgments about the possibilities it offers (Beck, 1992) and because as modern human beings we are all faced with huge ethical and ideological dilemmas, in an era in which our choices must be made in the absence of a widely accepted religious, secular or scientific ethic that can deliver concrete instructions to help us design our future (Beck-Gernsheim, 1996). However, this does not mean that in "advanced liberal" societies the body is finally privatized. On the contrary, following feminist

[5] See the works of N. Elias, P. Bourdieau or M. Foucault.

theory, as well as Foucault's approach, contemporary social theory commonly imagines the private body as itself a politically inscribed entity, shaped and marked by histories and practices of control (Bordo, 1999). Current discussions in medical sociology and anthropology, feminist studies and the sociology of the body, see bodies as concrete articulations of abstract social paradigms and understand modern societies to be regulating bodies, as part of their ongoing construction of a collective identity (Weiss, 2002). Extremely helpful for this line of thought, which connects between individual and collective bodies, are Mary Douglas' (1966, 1970) ideas about the material body as a receptor of social meaning and as a symbol of society. For Douglas, the so called individual-material-fleshly body is first and foremost a metaphor for society, or a conceptual microcosm of the "body politic". She thus sees the human body as perceived, interpreted and represented differently in different social times and settings, due to differing material cultures, technologies and means of control. Accordingly, the social system and cultural ideas about the body politic are reflected upon material bodies and the purity of the private and the national bodies are deeply interrelated. Since the social body shapes and constrains how the physical body is perceived and experienced, the sociological metaphor of the body is especially appropriate for locating "other" bodies, or the bodies excluded by society (Hazan, 2003). Wanted and unwanted fetuses for that matter are markers and reproducers of the body politic, just as their mothers, who are the biological "producers" of children and future citizens, are bearers not only of their own offspring, but simultaneously of collectives (Yuval-Davis, 1980). Having said that, it now remains to ask: what are the dictates of the body politic upon private bodies in Israel and Germany and how do they affect the field of reproductive genetics within both sites?

Concerning Germany, Herzog (1998) argues that in contemporary German public discourse, leftists and conservatives alike deploy holocaust images and memories in their political battles as a sledgehammer technique, in which invocation of the past and its crimes and shame became the lingua franca of post-war West German political culture. In accordance with this public atmosphere, regulation of the private body in Germany today is highly affected by the traumatic Nazi history and the Nazi regimes' handling of "unworthy lives", either for racial or for genetic reasons. This claim is exemplified for instance by Wuerth (1997). who argues that the history of the "Third Reich" is responsible for the unified German abortion law, which counterpoints between the moral bankruptcy of Nazi Germany, as well as that of former East Germany and between the moral, unified German state, which protects *all* life. This protection of all life, which is also manifested in Germany's position on stem cell research and in its embryo protection law, can be understood according to Wuerth, as a reaction to the past and as hallmark of the new "morally rehabilitated" Germany (see also Chapter 5).

Similarly, Linke in her book about German bodies writes concerning the German Green movement, that its fear of pollution, as well as its concern for the disabled and the mentally ill and the fight against vivisection, are based on an attitude which runs counter to the horrors of Nazism, where "mercy killing" was the fate of the disabled and of psychotics and where Nazi physicians were the founders of the logic and technique behind the final solution. Consequently, it is the aim of Germany today to fight for the rights of those populations that had previously been labeled "inferior". Nevertheless, Linke notes that the repertoire of symbols, images and metaphors, which are used as templates of protest, often curiously replicate the violent and murderous discourse of the past.

Concerning Israel, Hazan (1999) has claimed that the creation of the Israeli imagined community (Anderson, 1983) depends upon an intensive occupation with the body, as the Israeli-Jewish collective is symbolized in terms of its body. Equally, several writers have observed that the Zionist social movement has been extremely occupied with distancing itself from the iconic frail image of the body of the Diaspora Jew (Almog, 1997; Biale, 1986). However, by making this move a paradox emerged: while Zionist national discourse often perceives the existence of an independent Jewish state to be a consequence of the Jewish Holocaust, the similarity between the Zionist rejection of the Diaspora Jewish body and Nazi eugenic murderous practices are often overlooked. (Ivry, 2004; Kirsh, 2003). Even more so, Falk (2002), a prominent Israeli geneticist, reads the entire history of Zionism as itself a eugenicist project. Likewise, writing about excluded Israeli bodies, Weiss (1994, 2002) coined the term "the chosen body"(Weiss, 2002) to describe the Israeli collective body, which is a code according to which concrete bodies are monitored, screened, molded and selected from womb to tomb. According to her, the Israeli "chosen body" emanates from both the Zionist movement, which strove for the rehabilitation of the weak Jewish body, and the Jewish religious tradition, which is intolerant towards physical disability. Thus, the Israeli body of today worships healthy, fit, competent and whole bodies. Hence, Weiss argues that it is this "chosen body" collective ideal and its selectivity that is responsible for the Israeli quest for the "perfect child", a quest that by regulating mother's and fetuses' bodies also constructs the Israeli collective identity.

Nevertheless, in the following chapters, the relation of the present-day German body-politic to the Nazi past will be but one window through which I will understand German reproductive genetic culture, not the only or primary one. German attitudes towards science and progress and towards children and fertility, as well as the Christian and mostly Catholic, doctrine and its attitudes towards the beginning of life, suffering and the idea of "paying God", as well as Kantian philosophy, are to my mind very important factors shaping the German body-politic of today and its comprehension of the human body and the concept of "life" in all its forms and stages. Likewise, I will explain the selectivity characteristic of Israeli society by means other than those of Weiss. While Weiss's thesis centers on collective body ideals that lead to the quest for private and collective body perfection, and anti-disability sentiments also serve as a major explanatory factor in Remennick's thesis (2006) about why Israeli women opt for genetic testing during pregnancy, the explanation put forward in this book is more narrowly concerned with questions such as: what is the meaning of pregnancy in the Israeli-Jewish society? (Ivry, 2004), what is the difference between pre- and postnatal "life"? How important is fertility and what are the hegemonic conceptions concerning family responsibilities and the legitimate uses of technologies which manipulate life itself (Rose, 2001) Thus, for both Israel and Germany I argue for an analysis which does not focus mainly on rejection of the past (but rather on its diverse lingering effects) or on the future body of selected embryos and its suitability for national body ideals. Rather, my explanation centers on the more immediate meanings of the practices of reproductive genetics, or on the German and Israeli bio-cultural concept of the beginning of life and the way it responds to basic questions relevant to reproductive genetics such as: when does life begin? How thick is the border between pre- and postnatal life? When is life "worthy of living" and when is it "wrongful"? Moreover, how legitimate is the technological manipulation of life itself? However, the perceived "good society", in the sense of its genetic make-up and its relevance to the practice of repro-genetics is not overlooked, but rather postponed until the last chapters of the book and understood to result from a variety of complex factors. Having said that, let me now move a step forward and ask how national cultural repertoires concerning "the beginning of life" and the different Israeli and German collective bodies, effect genetic counselors' opinions, legislators decisions and the formation of the unique prenatal genetic scenes in the two nations.

THE EFFECT OF CULTURE

Up to this point I have claimed that experts' understanding of the human body and its "health", "pathology" or "viability" are dependent upon cultural/historical/religious/political categories and upon the complex relationship between the fleshly-private body and the body politic. However, how can we conceptualize the way in which culture shapes the practices of social agents such as genetic counselors or legislators deciding, for example, about "wrongful life" suits or about the moral and legal status of stem cells? Anthropological insights have shown that human beings live in webs of significance that they themselves have spun and continue to spin, as they make sense and attribute meaning to and for themselves. Those webs of significance are known collectively as culture (Geertz, 1973). Thus, culture is constitutive of human praxis, institutions and organizations, as it is the symbolic work we produce to reproduce ourselves. Traditionally, the explanatory power of "culture" has been conceptualized with concepts such as shared values or interests. These concepts overlooked the factor of social agency and perceived culture to be quite static and homogeneous. Later models have shifted attention toward conceptions of culture as affecting behavior by providing repertoires of practices, tastes, skills and capacities (Bourdieu, 1977, 1984; Swidler, 1986). Furthermore, newer models do not conceive of culture as static, reified, homogeneous phenomena, common to all members of national or ethnic collectivities. Rather, cultures are understood to be dynamic social processes operating in contested terrains in which different voices become more or less hegemonic in their offered interpretations of the world (Yuval-Davis, 1997). These newer theories, which break away from the concepts of values or norms, differ in the way they conceptualize the relationship between social agents and social structure and in their emphasis on questions of social power. Yet, what is common to most of them is the fact that they tend to overlook the relationship between cultural practices and moral reasoning. Because of this weakness and due to the centrality of moral questions related to the issue of "life", with which this study is dealing, I have chosen to work with the critical sociology of Boltanski and Thévenot (1999, 2006), which as argued by Resnik and Frenkel (2000), brings into the sociological discussion moral questions that have formerly been monopolized by moral philosophy. As a consequence, the innovation in Boltansky and Tavano's work is their use of regimes of justification in order to understand how culture shapes individuals' attitudes and actions. According to their basic assumption, humans are moral creatures who form their attitudes by adhering to shared metaphysical assets, which are offered to them by their culture. Thus, social agents employ moral reasoning and build on shared logics of justification: a limited number of "value worlds" that were historically constructed by their community and serve to form their attitudes, for justifying their own actions and for criticizing the actions of others. This way of thinking about culture, which stresses the centrality of moral justifications and sees moral discussion as an independent explanatory factor for social action, is especially appropriate for thinking about medical ethics, gene-ethics and questions concerning "(un)worthy lives".

As will be demonstrated throughout this book, experts do not form their practices in a vacuum, or solely out of their personal or professional experiences and expertise. Rather, they lean on local cultural logics of justification as they put to use cultural categories such as "life" itself as well as "normal" "healthy", or "wrongful" life. Moreover, the way counselors, parents, the disabled or legislators problematize different situations and the answers they seek and reach surrounding genetic interventions, cannot be understood as purely scientific, legal, economic (cost-efficiency considerations), or as the manifestation of interests. That is, since their practices always reflect a cultural-moral labor, which is required in order to interpret situations, to mutually adapt interpretations and to determine modes of agreement, which form a temporary moral-cultural order.

Of course, this process is not unilateral. Genetic counselors (as well as legislators), who are interpreters of knowledge and agents of biopower, also shape the logics of justification that are available in their society, in a bilateral process. Therefore, I will not try in this work to point out from where the practices I study initially emanate or to look for temporal causalities. Rather, I will provide a thick description (Geertz, 1973) of the culture of medical genetics in the two nations under study, as it is manifested in shared symbolic systems such as legal decisions, religious doctrine, public discussion, counselor's opinions, the opinions of the disabled and scientific *mentalité*.

To sum up: in this book I will show how scientific and technological knowledge is being played out against a background of national differences. In order to do that, I turn the focus to cultural logics of justification and to the moral discourse concerning the application of genetic methods during pregnancy in two nations, which is part of a larger question about the cultural meaning of "life" in our technological era.

However, this work will not overlook the structural characteristics of the field of reproductive genetics in the two nations. Looking from the perspective of culture, I understand cultural-moral reason has taken on a life of its own and become institutionalized as part of the different nation's exercise of biopower. Hence, this study does not distinguish between the spheres of ideas and institutions, or between politics, scientific innovations, law, economics and culture. Rather, it understands all institutions to be cultural and symbolic as well as part of society's structural constraints. Thus, following the tradition of cultural studies (Alexander, 1990), I will focus on cultural meaning systems as the driving force behind the differential adoption of reproductive genetics in Israel and Germany.

CHAPTER 2

METHODOLOGY

COMPARISONS

The comparative method used in this study aims at capturing both the general patterns of causalities or regularities across countries and the historical and contextual singularities. As a matter of fact, the type of differences I study emerge against a background of similarities between two modern, developed countries, with advanced scientific and medical systems and hence, with the same genetic knowledge (which serves as my "constant variable"). Although both are societies in which the field of reproduction is largely medicalized, a difference between Germany and Israel lies in the ways in which similar technologies are put to use and in the cultural logics of justification and rationalized myths built around them, as well as in their institutional strategies organizing fertility control. Hence, the comparison enables the study of how the meaning of biomedical technologies is assigned locally by practice and experience and not solely by their inner qualities. This kind of study is important because social practices tend to become institutionalized and reified (Berger and Luckman, 1967) and, as a result of this process, moral debate, which is the focus of this study, is blocked. The comparative analysis applied in this work helps to fight this tendency by taking a fresh look at what, from within specific cultures, seem to be solid practices that offer no alternatives. This results from the fact that the comparative method builds on the assumption that cultural differences help to unveil the taken for granted social order (Lamont, 1992), since by looking from the perspectives of two cultures, the researcher is supplied with vantage points that enable the estrangement of social attitudes and practices. Consequently, the reciprocal gaze from one culture to the other allows for an examination that is less bounded by cultural idiosyncrasies. Accordingly, in this work I attempt to shift back and forth between the two scenes of study without favoring the perspective of either of them and to use the differences and similarities found as analytical tools, which stimulate thought (Ivry, 2004).

In order to compare the contemporary reproductive genetic scenes of Israel and Germany, I used both qualitative and quantitative methods, as well as textual analysis.

THE QUALITATIVE DATA AND SAMPLE

The empirical quantitative data presented in this study come from a study of Israeli and German genetic counselors' professional practices. The samples are based on data collected through questionnaires (N = 295), which were distributed to genetic counselors in Israel and Germany in 1993–95 and 2000–01. The first pool of 207 questionnaires comes from data collected by Wertz and Fletcher in 1993–95 as part of their cross-national study of 37 countries (Wertz and Fletcher, 2004). The original questionnaire included fifty questions dealing with ethical and social issues in genetic counseling. It comprised demographic and professional questions (such as place of training, professional education or sub-specialization, years of experience on the job, work in public or private setting, etc.), case vignettes describing possible scenarios in the clinic, which concretize the issues in question and different closed-ended questions devised to reflect on a broad range of issues, such as selective abortions in the case of different genetic conditions, perceptions of disability, understanding of the state's role regarding reproductive genetics and the like. (See also Wertz and Fletcher, 2004).

The late Dorothy Wertz allowed me to use their raw data, from which I extracted and re-coded this study's first database. The second database, which corresponds to the first one, contains the answers to a questionnaire that replicates parts of the original study and which I distributed and collected in both countries in 2000–01 (N = 88).

The First Survey

In Wertz's study, in each of the 37 surveyed nations, a local geneticist compiled a list of practising geneticists and distributed and collected the anonymously answered questionnaires. In Germany,[1] Irmgard Nippert and Gerhard Wolff distributed the questionnaires to professional geneticists, based on the membership lists of the German Society of Human Genetics (*Deutsche Gesellschaft für Humangenetik*, also known as the Scientific Organization of Geneticists), the Professional Board of Geneticists (*Berufsverband Medizinisce Genetik*) and listings from all the genetics clinics and academic departments in the country. A German translation of the questionnaire was used.

In Israel, questionnaires were distributed by Zully Kohan, though only to MD geneticists (and not also to non-MD genetic counselors). Both Hebrew and English questionnaires were used. The number of Israeli MD counselors at that time was very small (around 30) and so Zully Kohan could distribute the questionnaires to all of them based on her acquaintance with them.

Unlike this study, Wertz's international survey did not focus only on practices of prenatal diagnosis but rather on a broader range of issues in genetics and ethics. Additionally, it included non-counseling geneticists among its research population. In order to prepare the data for use in this study, only those questions relevant to the specific concerns of the current research were analyzed (15 out of the 50 original questions). Moreover, in order to learn specifically about practices of counselors who advise patients, only practitioners of genetic counseling were included in the survey. Twenty-seven Israelis were invited to participate in the original study, of

---

[1] In Germany, the study was funded by the *Deutsche Forschungsgemeinschaft*. Internationally, the study was supported by the Ethical, Legal and Social Issues (ELSI) program of the U.S. Human Genome Project (National Institutes of Health).

whom 23 (85%) responded. I used only 22, on the basis that they consulted with at least 1–5 patients per week. In Germany, 418 geneticists were invited to participate and 255 (61%) responded. Of these, I used only 185, again on the basis of seeing at least 1–5 patients per week.

The Second Survey

In order to look for the influence of time on genetic counselors' practices and to check for the possible effects of the rapid developments in the field of human genetics on professional norms in both countries, in 2000–01, I replicated parts of Wertz's study and distributed a follow-up questionnaire based on it, thereby collecting new data focused on reproductive genetics (see Appendix 6). Respondents in the second survey might also have responded to the first one but the two populations did not overlap. The completion of the new questionnaire took up to one hour. In Israel, the questionnaires were distributed in English; in Germany, in German. This was due to the fact that Israeli geneticists' knowledge of English is very good, as it is their professional lingua franca. In Germany, knowledge of English is more limited, as scientific education and careers can be fulfilled in German.

In Israel, the anonymous questionnaires were distributed in several ways: some were handed out at professional meetings in which I participated, while others were distributed by post to the mailing lists of two organizations: The Israeli Association of Medical Geneticists and The Israeli Association of Clinical Geneticists (Non-MD genetic counselors). The rest were sent to employees of genetic hospital units, after receiving permission from the head of the unit. After the questionnaires had been distributed, participants were contacted by telephone and asked to return them. Forty-four Israeli genetic counselors (this time including non-MDs) were invited to participate in the study (at the time of the research, the total population of genetic counselors included about 55 people). Thirty-one of the 44 contacted counselors (70%) responded. In Germany, 161 questionnaires were mailed (in double envelopes) all over the country to counselors belonging to the German Professional Board of Geneticists (*Berufsverband Medizinische Genetik* and a list of genetic counselors: *Genetische Beratung in Klinik und Praxis*).[2] Only a few – those whose email address appeared on the list – were contacted a few weeks later by email and reminded to fill in the questionnaire (phone calls were not made due to budget limitations). Four addressees returned empty questionnaires,

[2] According to Christina Scholz, head of the BV main office in Munich, there are no programmatic differences between the two societies of geneticists: the German Society of Human Genetics (*Deutsche Gesellschaft für Humangenetik*), also known as the Scientific Organization of Geneticists, and the Professional Board of Geneticists (*Berufsverband Medizinische Genetik*), which is a professional organization. Seventy-five percent of the BV members are also members of the German Society for Human Genetics and 50% of the members of the Society of Human Genetics are members of the BV.
According to interview materials, in the past, geneticists who work in hospitals dominated the BV, while the Society for Human Genetics was dominated by private practitioners. However, this has changed. In Israel most counselors work in hospital-based clinics.

explaining that they could not answer it for various reasons, such as having left the field of counseling in favor of laboratory work. Fifty-seven questionnaires (35%) were returned. I cannot estimate how many people might not have received the questionnaire or found them to be irrelevant to their line of work, since no contact was made other than through the above-mentioned list, which unfortunately is updated somewhat infrequently. Counselors who had moved without leaving a forwarding address, had passed away or had left the field, might thus reduce the original sample considerably, in both the first and second surveys.

Possible Biases of the Second Survey

In Israel, most of the unreturned questionnaires were from two hospital genetic institutes. Based on interview materials, I have reason to believe that these institutions are very permissive in their attitudes towards the selection of embryos. This stance could be a possible explanation for their unwillingness to participate in a study inquiring into moral practices. It follows that the Israeli sample might suffer from a degree of underestimation of counselors' permissiveness in selectively aborting abnormal fetuses. On the other hand, all contacted German counselors were aware of the fact that this study was being conducted by an Israeli (and so probably Jewish) sociologist. As the topic of embryo selection in Germany is highly related in professional and public opinion with Nazi eugenics (Cohen et al., 1997; Erikson, 2003; Krones, 2005; Krones and Richter, 2004; Nippert and Wolff, 2004), the fact that they were approached by an Israeli might have affected their answers or indeed their willingness to participate at all. This might lead to an effect of overestimation of the phenomenon in the German case; in other words, German counselors might seem to be more cautious with the practice of PND and its outcomes than they really are. Moreover, there are inherent, unavoidable weaknesses in the questionnaire method, no matter how carefully contrived, such as the inability of a questionnaire to describe complex cultural logics of justification. Therefore, in-depth semi-structured interviews were also conducted.

THE INTERVIEWS

In the years 2000–2001, I conducted 32 in-depth semi-structured interviews with Israeli and German genetic counselors: 18 in Israel and 14 in Germany. I first approached Israeli counselors in a meeting of their organization (clinical rounds), while others were reached through membership lists of their organizations (The Israeli Organization of Medical Geneticists and The Israeli Organization of Clinical Geneticists), or following a referral from a colleague (snowball sample). In Israel, interviews took place in eight out of the 11 active hospital genetic institutes, all over the country (Haifa, Tel-Aviv, Beer-Sheva, Rehovot, Jerusalem, Hulun, Kfar-Saba, Ramat-Gan). Among the interviewees were seven heads of such institutes. Most of the interviewees were M.D. geneticists who specialized in diverse medical fields such as pediatrics, gynecology and internal medicine. Six of the interviewees were non-M.D. counselors. An extra non-anonymous interview was conducted with Prof.

Joël Zlotogora, Head of the department of community genetics in the Israeli health ministry, (Prof. Zlotogora renamed this unit, which was formerly called "the unit for the prevention of birth defects"). All interviews were conducted in Hebrew and most interviewees had formerly filled out the questionnaire.

Due to the large size of the country and because of budget limitations, interviews in Germany were mostly conducted in the province of North Rhein-Westphalia (Nordrhein-Westfahlen). Using the list of the German professional board of geneticists (Berufsverband Medizinische Genetik, BV), I contacted all counselors who were working within two hours driving distance from the city of Cologne, where I was staying. This included other towns like Essen, Düsseldorf, Düren, Aachen, Bonn and, of course Cologne itself. Among my German interviewees were counselors holding diverse positions and educational backgrounds: genetic counselors who are gynecologists and pediatricians, four hospital unit directors, a counselor working in a Catholic hospital, a former president of the German Society for Human Genetics, a university professor in the field of genetics who also practices counseling and formulators of guidelines in the field of human genetics. In Germany, because the size of the population I was studying is much larger than in Israel and because of a lower response rate, only a few of the counselors who were willing to be interviewed, also formerly returned the questionnaire by mail. However, when they were contacted by phone from within Germany, all counselors in the area invited me for an interview.

The interviews took place in hospital units, university departments or university-based hospital institutes, as well as in private practices and homes (7 out of 14). The large proportion of private practitioners presented in the interview pool has to do with the occupational structure of genetic counselors in Germany, which will be discussed in Chapter four. An extra interview was conducted at the University of Münster with Prof. Irmgard Nippert, who specializes in research concerning human genetics in Germany.

The interviews were conducted in order to reach the complex logics of justification that genetic counselors use in their value-laden field of work or, in other words, to look for the "value worlds" that were historically constructed by their community, and which serve them in forming their attitudes, justifying their own actions and criticizing others (Boltanski and Thévenot, 1999, 2006; Resnik and Frenkel, 2000).

The interview questions were developed after a review of the literature on the subject and a preliminary analysis of the quantitative materials. When setting out to an interview I had prepared a set of fixed questions (see Appendix 6) about opinions concerning different issues, such as the standards for entry into the human community, the good of society, the legal and moral standing of the fetus and its parents, disability, the goals of genetic counseling, science and progress and the history and consequences of medical genetics.

However, in practice the conversations were open-ended. I set out with one set of questions but enlarged and transformed them as during the interviews I became more knowledgeable about my field of study and its dilemmas. In order to learn more about the reality of their work, I asked the counselors to describe problematic (and non-problematic) counseling sessions, in which abstract moral values were confronted with mundane professional decision making and with parents' fears and sorrow. In addition to questioning about professional practices, the interviews were also used to learn about the institutional background of genetic services in both countries, i.e., about the legal situation, profit motivations, peer pressures and the like. According to my interview technique, I tried to be non-directive in the sense of not pushing towards a "desired" answer but this did not mean being silent or not sharing information, opinions and emotions. Especially in Germany, I sometimes used examples from the Israeli situation in order for the counselors to feel less obliged to present me with the "official" reasoning common to their field. By referring to comparative data, I tried to make myself more reliable, and to free the interviewees from the burden of guilt feelings towards a Jewish researcher. In doing so, I tried to avoid the effect of our common tragic history, which might have caused the answers given to be slanted in the direction of presenting me with what are considered to be high moral standards and not with questioning them.

After the formal interviews, interviewees (especially women counselors) shared with me their private experiences as both , experts and pregnant women at the same time. In those cases, we shared our experiences together and I lacked neutrality, as I identified very strongly with the fears, comfort, and mostly with the enormous responsibility prenatal genetic diagnosis bestows upon women. In Germany, many interviewees were eager to hear about genetics in Israel and about the political situation in my home country, which was worsening day by day in the summer of 2001. My German-Jewish background and my feelings towards modern Germany were other subjects that seemed to reappear during the interviews, especially with older counselors who carried personal memories from WWII.

Another subject that was often discussed in the German interviews was stem cell research and the connections between Israel and Germany concerning this field of study. Just before my arrival in Germany, the German media had extensively covered a story about a German geneticist from the state in which I was living and working, who imported human stem cells for his studies from an Israeli colleague. In importing the human stem cells, the German scientist used a loophole in the German law of that time, which prohibited research with German human stem cells but did not forbid their importation from other countries where human stem cell research is allowed, such as Israel. This story had caused a lot of controversy in Germany about research with imported human stem cells (Jewish ones!), which was followed by an incisive discussion among German politicians and public figures (Prainsack, 2006a). In the interviews, German counselors repeatedly brought up this subject, although it belonged only to the outskirts of my research interests. Yet, it does illuminate once again the extreme caution with which genetic matters are treated in Germany, as opposed to the seemingly "non-issue" treatment they receive in Israel's public opinion, laws, politics and media, which to my great surprise, did not cover the German-Israeli stem cell story at all.

Interviews in both countries lasted from 1.5 to 4 hours. All interviews were tape-recorded, as well as summarized by taking written notes and all interviewees were promised anonymity. The interviews in Germany were conducted in English. Whenever a language problem appeared, I was assisted by a native speaker.[3] Interview data were coded and analyzed thematically.

No matter how many people I interviewed, there was always someone who told me a different story. Furthermore, despite the significant differences found in the quantitative analysis between German and Israeli counselors and despite the national differences, which were clearly represented in the interviews, there was always a possibility of two counselors from two different nations sharing more opinions with one another than with their colleagues at home. Yet, in general, the variance that was found between the national groups was larger than the differences within the groups.

This, of course, does not mean I claim complete homogeneity among members of different national cultures, as structures of meaning underlying any social situation are always multiple, partial and tangled together.

Anonymity

The interviewees who were active counselors were promised anonymity. Only Prof. Zlotogora, head of the department of community genetics in the Israeli health ministry is mentioned by his full name. The Israeli population of counselors is relatively small, and almost all its members know each other. In that sense, anonymity for people who discussed difficult ethical situations and personal values with me, is hard to provide, since by giving even the most general characteristics of the interviewee, s/he could easily be identified by peers. Therefore, when quoting from interviews I will sometimes use false identifications, or very general ones.

Also in Germany, because the geographical area in which the study had been conducted is limited, the problem of providing anonymity exists. That is because geneticists working in the same federal state tend to know each other. Therefore, counselors will only be vaguely described in this work.

Possible biases: in Israel, my request for an interview was turned down by two genetic institute managers. According to information collected through other interviews, I have reason to believe that these institutions are especially permissive in their attitudes towards the selection of embryos (as they are the same ones alluded to above, who also did not return the questionnaires). This stance could be a possible explanation for their unwillingness to participate in a study inquiring into ethical positions. For this reason, the Israeli sample might suffer from some underestimation of the phenomenon of the Israeli counselors' support for the detection and selection of "abnormal" fetuses. On the contrary, all contacted German counselors were aware of the fact this study was conducted by an Israeli (Jewish) sociologist. As

[3] My assistant Marcus Paus, who has an excellent knowledge of both German and English and who, as a medical student, also has a good understanding of the research topic.

the topic of embryo selection in Germany is highly related in the professional and public opinion with the Nazi eugenic programs, the fact they had been approached by an Israeli, might have affected their willingness to participate, as well as their answers. This might lead to an effect of overestimation of the phenomenon in the German case, which has caused German counselors to seem more cautious with the practice of prenatal diagnosis and its outcomes, than they really are. Having based this work also on quantitative data collected by different people, as well as on textual analysis, will hopefully reduce the dangers of such a bias.

In Germany, the fact that I was interviewing in only one federal state might cause a slight regional bias. However, North Rhein-Westphalia is not known to be extremely liberal (like Hamburg, for instance) or conservative (like Bayern). Furthermore, taking into consideration the fact that questionnaires were collected from all over Germany, I assume that the sample is fairly representative.

Only one of the counselors I called up while working in Cologne could not be interviewed because of my poor German and her poor English. Most non-English speaking counselors in Germany have been educated in the former East Germany, or in other East European countries. This, and the fact that I was living in what used to be West Germany, has rendered this relatively small group of former East German counselors hardly present in my interviews.[4] As East Germans are known to be more secular, more accepting of abortion and also more directive in their counseling (Cohen, Wertz, Nippert and Wolff, 1997), this difference could not be accounted for in my research.

THE TEXTS

Apart from collecting quantitative data and interviewing, this study is also based on analyses of texts. The majority of the textual materials studied in this work are legal ones such as abortion laws, embryo protection law and court decisions concerning "wrongful life" suits. Another group of analyzed texts is the professional writings of ethics committees (for example the German National Ethics Council, 2003) and of medical geneticists, among them: professional guidelines and policy recommendations, conference lectures and discussions and the like. Furthermore, the views of disability organizations concerning prenatal genetics were collected.

Newspaper materials were also collected. In Israel, during the 5 years of my research I collected every mention of reproductive genetics, either in newspapers or the electronic media. Whenever I missed out on some reports, there were always colleagues, friends and family members familiar with the topic of my research who guided me to the relevant materials. Thus, although the collection of materials

---

[4] In the mid 90s, East German counselors made up about 20% of the German geneticists population, (See Cohen, Wertz, Nippert and Wolff, 1997).

was not done very systematically, I believe it encompasses the majority of media materials dealing with reproductive genetics in the mainstream Hebrew Israeli media.

In Germany, the majority of newspaper articles collected for this study were texts published in the *Frankfurter Allgemeine Zeitung* (FAZ),[5] English edition, through the years 2000–2001. Studying the German FAZ was done more rigorously, as I used the newspaper's Internet archive to look for articles including such words as: prenatal diagnosis, wrongful life, abortion law, embryopathic indication for abortion, eugenics, stem cells, PGD and the like.

WHO IS NOT REPRESENTED IN THIS STUDY?

It is important to point out that my work focuses on Israeli-Jewish culture and not on Palestinians either living in Israel or in the Palestinian authority. That is due to different reasons. The most obvious one is that focusing on a professional group (and not on the general public) resulted in interviewing only Israeli-Jewish counselors, as at the time of the research there were no Palestinian counselors working in Israel. Secondly, state policies and laws regarding reproductive technologies are constructed with regard to Jewish-Halakhic concerns (Kahn, 2000) and not to Muslim concerns. Thus, it should be clear that when I refer to the "Israeli" scene of reproductive genetics or to the "Israeli society" I am referring to Israeli-Jewish society. Likewise, in Germany none of my interviewees belonged to a minority group and all were either Christian or non-religious. Additionally, my research was also not designed to cope with differences that might be found between former East and West Germans. Thus, while making generalizations about Israeli and German counselors or cultures, I unavoidably tone down individual differences or even tensions present within each society. This, however, does not mean I believe either the German or the Israeli cultures to be homogenous entities. Rather, it results from the limitations of a comparative framework that centers on broad national differences and similarities and not on contentions within each culture.

The empirical study was completed in 2003. I have since been updating parts of the data, especially those concerning regulations relevant to the field of reproductive genetics in both the studied societies. However, changes taking place after the end of the study or after handing in the manuscript for printing, may not be represented in the book.

---

[5] The FAZ is published in Frankfurt but read all over Germany and internationally. It is considered to be a highbrow newspaper, targeting an educated audience. The FAZ focuses on politics, financial and business matters but also on culture, literature and science. It is published daily, and is known for its excellent Wednesday science supplement. The FAZ is considered to be quite a conservative newspaper.

CHAPTER 3

GETTING TO KNOW THE FIELD OF REPRODUCTIVE GENETICS IN ISRAEL AND GERMANY

The power of modern societies over life and death breaks down in everyday life to many different structural and institutional opportunities and constraints. Hence, when writing about a medical innovation concerned with the balance between the quality and sanctity of "life", which enters a national medical market, it is obviously important to notice the historical, institutional, legal, political, cultural and religious contexts in which this process takes place. For that reason, the following chapter is dedicated to a comparative overview of the Israeli and German fields of prenatal genetics.

In it, I will discuss different actors involved in the field of prenatal genetics in Israel and Germany, among them: the field's experts, the field's supporters and opponents, the field's earners, such as private and public clinics that offer prenatal genetic tests and above all, the institutional landscapes in Germany and Israel, with their unique health care systems, history and support of medical genetics. My leading question will be: what is it about the field's history, contemporary political and financial organization and cultural surroundings that can shed light on the dramatic differences found between Israel and Germany in their adoption of genetic reproductive practices.

As in the rest of this book, in this chapter, I will also not employ a sharp distinction between the fundamental concepts and ideas of medical genetics and its institutions. Socio-historical and cultural factors take on a life of their own, become institutionalized and form the world of prenatal medical genetics in which moral reasoning, budgeting and institutional arrangements work together and mutually affect one another in various ways. Thus, before getting to know how this field is organized today and the contemporary opportunities and constraints embodied in it, let us look back in history, in order to learn how it became the way it is and what was the process of its institutionalization.

A HISTORICAL PERSPECTIVE

Human Genetics after the War in Germany

In the Nazi era, prominent German human geneticists had actively participated in formulating and implementing Nazi race ideology. Under the pretext of eugenically efficient measures, the Nazi party enforced sterilization, termination of pregnancy, asylum, prohibition of marriage and the murder of patients. Consequently, the Nazi racist ideology with its eugenic roots led to the declaration of Jews as foreign genetic material and this definition paved the way for their genocide (Müller-Hill, 1984).

However, even before the actual extermination of Jews took place, the two most notorious eugenic measures taken by the state were a eugenic sterilization law and the euthanasia program, which targeted the general German population. The 1933 sterilization law forced compulsory sterilization for a variety of medical conditions that were thought to be genetic in origin such as: congenital feeble mindedness, schizophrenia, manic-depressive insanity, genetic epilepsy, Huntington disease, hereditary deafness, hereditary blindness and heavy alcoholism. Between 1934 and 1939 it is estimated that about 350,000 Germans were sterilized. Later on, between 1939 and 1941 within the framework of the "mercy killings" of the "euthanasia" program, over 70,000 children and adults, whose lives were declared "not worth living"(Müller-Hill, 1984), were murdered. When this program was stopped due to public pressures, the gas chambers, which were used for the first time in this enterprise, were transported with their operators to Poland, to serve in the "final solution". Needless to say, one form of murder led to another (Bock, 1984; Friedlander, 1995). And yet, one must remember that eugenic ideology and measures were in no way a German particularity (Adams, 1990; Kevles, 1985; Koch, 2004; Paul, 1995, 1998;). According to Müller-Hill (1996), Germany in the 1930s was seen as the avant-garde of eugenics, as almost all non-German eugenicists praised the German sterilization law of 1933. Accordingly, it was the violent anti-semitism and not the racial hygiene, which could not be sold to the international community. Even when the war was over, the sterilization law was not immediately perceived as a typical Nazi injustice and it was only in the 1990s that the surviving sterilized individuals were finally regarded as victims of the Nazis. Thus, as we can see, inside Germany after the war, human genetics was not very badly discredited because of its past. Rather, the revival of genetics after the war in Germany was characterized by a gigantic repression of its history (Müller-Hill, 1987). At that time, there was no discussion about the odious past in the various German journals that covered this field (Müller-Hill, 1996) and even as late as 1958, in a meeting of the Society of Human Geneticists, the speakers included former Nazi scientists. Among them were, for example, an expert from the SS for "cleaning" the population, a geneticist who worked in Auschwitz, a student and collaborator of Dr. Mengele and the former head of the "*Rasse und Siedlungshauptamt*" (main office for race and settlement) of the SS. Nevertheless, genetics was lagging behind in Germany, compared to other advanced societies, up to the 1970s. Thus, Deichmann (1996, 2002) asks what, if not bad reputation, caused the slow start of molecular genetics in Germany after the war? Looking into this subject, she argues that neither National Socialist research policy, nor the expulsion of Jewish scientists, despite the significant losses they entailed, can be used as the sole explanation for the lag in molecular genetics research in West Germany up to the 1970s (Deichmann, 2002). Other explanations such as the closing of some universities during the war, or the destruction and money shortages after the war ended, also cannot suffice. Otherwise, the misunderstanding of the importance of this field in post-war West Germany can better be explained according to the following reasoning: firstly, the politically conditioned lack of international scientific exchange with Germany after the war contributed to the fact that German science did not catch up with the international developments in genetic research and technologies. The isolation and self-isolation of German scientists reached its peak after WWI, and yet, also after WWII German scientists encountered a lot of mistrust coming from the international scientific community. This is despite the fact that in contrast to the situation after WWI, the Cold War pushed towards integrating Germany into the West as quickly as possible, politically, economically and also scientifically. Nevertheless, according to Deichmann, even a partial isolation of German scientists was of importance to the development of molecular genetics. Analyzing correspondence between geneticists after the war, she reveals that many of the foreign scientists, Jews and non-Jews, Europeans and Americans, were unwilling to meet with German colleagues for years, at least those from the older generation, who were (rightly) suspected of having collaborated with the Nazi regime. For example, even 18 years after the war, a discussion about the location of the 1963 International Genetics Congress reveals that Germany was still not considered a neutral option. In response to an anti-Semitic act in the Federal Republic of Germany in 1960, it became clear that not only Israeli, but also a number of American and European geneticists were not yet ready to attend a congress on German soil. The resistance had also to do with the fact that former Nazi geneticists were once again occupying prominent scientific positions in post-war Germany. Apparently, history and the Nazi regime's use of a pseudo-genetic reasoning for the mass destruction of human beings, with extremely little opposition from German scientists, could not be so easily forgotten by the international community. And so, in this specific case, the location of the congress was changed. Deichmann's second explanation for the slow start of molecular genetics in post-WWII Germany has to do with the expulsion of Jewish scientists. This not only led to the expulsion of great scientists, some of whom contributed decisively to the field of molecular biology, but also to a decline of "dynamic biochemistry", a field which later proved to be important for molecular biology. All this resulted in a situation in which at West German universities, until 1961, there were no genetic institutes in the Natural Sciences Faculties. Additionally, after the war, human genetics as a scientific discipline was scarcely represented in medical school curricula (Nippert, 1998). The effects of this lack of medical genetic education can still be felt today. Writing for a special supplement of the European Journal of Human Genetics, comparing medical genetics in 31 countries, Harris and Reid mention that in Germany, in the late 90s, access to genetic services was still limited by doctors' lack of genetic knowledge (Harris and Reid, 1997). Furthermore, in 1987, a *Bundestag* committee appointed in order to discuss chances and risks of gene technology (*Bericht der Enquete-Komission "Chancen und Risiken der Gentechnologie*", 1987) mentioned a shortage of genetic counselors in Germany. Similarly, in my own interviews conducted in the summer of 2001, heads of genetic institutes and clinics, private and public, complained about a lack of younger-generation geneticists.

Human Genetics in Germany from the 1960s onwards

In the 1960s and early 1970s only a few departments of human genetics offered genetic counseling, while counseling centers did not exist at all in Germany. In those days, new genetic screening technologies were hardly known or available to the public and it was also illegal to terminate a pregnancy for genetic reasons or fetal abnormality (Nippert, 1998; see also a discussion about abortion laws in Chapter 5). It was in 1972 that the most prominent step in the introduction, implementation and diffusion of prenatal diagnosis (PND) took place in West Germany. At that time, the German Research Foundation (*Deutsche Forschungsgemeinschaft*) decided to award a multi-million DM grant for a seven-year multicenter collaborative study on the diagnosis of genetic defects (Nippert, 1992a). In addition, in June 1973, a wealthy member of the main organization of the mentally handicapped ("*Lebenshilfe*") governing board made a 350,000 DM contribution in support of clinical application research of PND.[1] The donation provided for the foundation of the first genetic counseling center in Germany, which was located in Bavaria. Thus, in 1975, Bavaria also became the first German state to provide public funding for PND. By the mid-1970s, genetic counseling centers offering genetic services and PND were established at universities all over Germany and amniocentesis (AC) was implemented in prenatal care via genetic service provision. The implementation of the centers at universities is due to the fact that in those days the special skills required for genetic counseling were scarce and there was also no profit motive for practitioners in this field (Nippert, 1998). Having the federal states fund an out-patient medical service like genetic counseling in their medical schools, as was the case with the university genetic centers, was unusual in the German health-care system, in which out-patient services are typically delivered by private physicians. (This unique arrangement still has its effects nowadays, as will be explained below under the heading of profit motivations). Other factors attesting to the steady increase in PND from the mid 70s to the 80s, are the following structural changes, which made PND services more accessible and available (Nippert, 1992a):

1. Since 1975 costs related to PND have been covered by health insurance
2. Since 1976 termination of pregnancy due to a eugenic indication has been allowed up to the 22nd week of gestation.
3. Direct state government financial support (mainly of state universities) allowed the expansion of tests and counseling.

The field was shaped during those years and amniocentesis (AC) became a standard part of pregnancy quality care in Germany, especially with the indication of advanced age of the mother-to-be. Concurrently, German human geneticists began to imitate their American colleagues in treating their patients, as they adopted new concepts relating to the doctor-patient relationship such as: informed consent, patient's autonomy and non-directive counseling. It is also ever since then that PND has remained a special service at university level, although today it is also widely practiced privately.

After 1948 (the formation of the state) in Israel

A variety of human genetic studies took place in the modern state of Israel immediately upon its establishment (Goodman, 1989). The main impetus for this initial flurry of activity came from the massive number of Jewish immigrants arriving in

---

[1] This organization later opposed PND.

Israel from all parts of the world, a process which led to the perception of Israel as an ideal setting for studying genetic similarities and differences among the Jewish population (Kirsh, 2003). As a result of this scholarly effort, it was soon recognized that frequencies of genetic diseases differed markedly among the various Jewish communities, a finding which further served as a catalyst for studying the genetic make-up of these groups (Kirsh, 2003). Therefore, most of those early Israeli studies in population genetics tried to answer questions such as: How heterogeneous are the different Jewish communities? How much do they differ from one another and from their former host populations? and To what extent do the genetic data correlate with the known histories of the separate groups (Kirsh, 2003)? These lines of research went hand-in-hand with efforts to apply the knowledge they produced, and so, according to Cohen (1992), doctors and scientists in Israel began providing genetic counseling as well as treating and investigating hereditary diseases long before medical genetics was recognized as a medical field. Consequently, as early as 1964, the first department of human genetics and a genetic counseling clinic was opened in Jerusalem, through the initiative of Prof. E. Goldschmidt and colleagues at the Hebrew University of Jerusalem. Furthermore, research into population genetics in Israel during the 50s and 60s was a field of knowledge that was considerably affected by historical processes, ideology and socially-determined perceptions (Kirsh, 2003). Hence, Kirsh claims, a clear effort was being made by the scientists to shape and ratify the emergent Jewish national identity by genetically proving a shared biological origin for all Israelis, sometimes at the cost of somewhat biased research conclusions. And so, the studies done in those days proved that there was an absence of substantial gene influx from non-Jewish populations into the Jewish gene pool and some of their highlights were findings showing that genetic distances among most Jewish communities from Europe, Africa, and Asia were smaller in comparison to those between Jews and the indigenous populations of their country of origin (Cohen, 1992). Trying to look for a common origin for the major Jewish ethnic groups in a context outside the Jewish state might have been perceived as using medical genetics to ask racist questions. However, this is not how it was seen in Israel and even today, similar studies into the "Jewish Genome" are being carried out (Hammer et al., 2000; Thomas et al., 2000). Indeed, Rafael Falk (2002), a prominent Israeli genetics professor, reads the entire history of Zionism as a eugenicist project. He states that the understanding of Judaism as a biological essence became an integral part of Zionist thought towards the end of the 19th century. While most European Jews tried to fight against the idea of Judaism being a "race", prominent Zionists such as Hess, Herzl, Bialik, Nordau and even Buber argued that the biological dimension of the Jewish "*Volk*" should not be overlooked. Similarly to Falk, Stoler-Liss, who writes about Zionist motherhood, also claims that in the 20s, 30s and even beyond, eugenic thought was prevalent among Zionist pediatricians, gynecologists, general practitioners and other types of experts and "advisors" in the Yishuv[2], who were trying to both quantitatively and qualitatively better future generations of "*Zabarim*" (Stoler-Liss, 1998, see also Hirsch, forthcoming).

[2] The Jewish community in Palestine before the formation of the Israeli state in 1948.

The Lasting Effect of the History of Eugenics in Both Nations

Whereas in Germany the proximity of their murderous eugenic past affected the field of medical genetics after the war in a host of ways, this European history with its fatal effects for European Jews, went unnoticed in the early days of the Israeli state (Kirsh, 2003). Likewise, my interviews with Israeli human geneticists, which took place over half a century after the end of the war, show that they perceive themselves to be the victims of racism (as opposed to the descendants of persecutors of racist atrocities) and hence they do not pause for moral contemplation about their professional activities and their possible relatedness to wrongdoings of the past. Moreover, for most of the Israeli public and the vast majority of Israeli professionals (with the exception of a few critical thinkers), the kind of eugenics that was condemned in the past is seen to bear no relation whatsoever to contemporary practices. This can be demonstrated for example by the writings of Chemke and Steinberg (1989), two prominent Israeli human geneticists, about the cultural context of medical genetics in Israel. While they mention the fact that the renewed state of Israel was built "*upon the most tragic event in human history, the Holocaust, during which 6 million Jews were killed...*" and that "*...There is an obviously strong memory of the terrible events only half a century ago, creating a strong desire for survival*" (p. 280), this strong desire for survival, or the memory of the Jewish genocide, is not related to medical genetics in the past or present. Rather, the past is mentioned as creating a desire for survival and for becoming stronger, but not as leading to any moral contemplation. On the contrary, fear of the revival of eugenic policies is often mentioned in writings about contemporary reproductive genetics in Germany (Erikson, 2003; Krones and Richter, 2004; Nippert, 1992b), as guilt and shame play a large role in current German politics. Moreover, Cohen et al. (1997) mention that the reaction to the experience of the "Third Reich" has even been manifested by violent threats and actual bomb attacks on prenatal diagnostic laboratories and German counseling centers. Likewise, German geneticists have been called "the new eugenicists" by different political, feminist and disability rights groups (Schroeder-Kurth, 1990).

These different atmospheres clearly appeared in the interviews I conducted. Whereas in Germany, not one of my interviewees treated accusations saying that nowadays genetic counseling is basically a contemporary form of eugenics, lightly; in Israel, similar accusations were regarded as cynical by most counselors. For example, the head of a genetic institute in a hospital nearby Tel-Aviv told me:

> *"Those accusations make me laugh. When my grown-up children try to tease me, they say I am practicing eugenics. Of course, I do not take this seriously".*

What is interesting, is that this non-critical thinking about medical genetics has also been found to characterize Israeli disability activists (Raz, 2004), in sharp contrast to their German counterparts (see Chapter 7).

In Germany, when I asked about the influence of the past, most interviewed counselors felt very strongly that it must never be forgotten and that it should serve as a constant warning for their own current actions. Yet, some German counselors also saw the repetitive mentioning of the past as a trend that blocks fresh moral thought. For example, an MD human geneticist, a woman working in the private sector in Germany said:"*Because of our history, I am afraid that we think of the moral dilemmas that have to do with reproductive genetics in an old fashioned manner, like people in the West thought of the mini-skirt in the 70s... Are we better than the Israelis or the Americans? I don't think so, but it is very important for us to seem better because of our collective guilt feelings. This leads to a group dynamics that blocks moral thought and to moral hypocrisy. I personally feel that it is very difficult for me to state a non-official opinion that does not have to do with high moral principles but with the difficult reality of human life"*

Whether being critical of the influence of history or not, the past was clearly alive and kicking for all my German interviewees and far less so for their Israeli counterparts. The effects of history are thus as follows: Israeli counselors have a double eugenic memory of both the atrocities of the Nazis and the Zionist "soft" eugenic history. Whereas the first memory has been repressed, forgotten and understood as irrelevant to today's practices, the second, namely the not criticized Zionist-Jewish" "soft" eugenic history is alive and well, although it is also never explicitly connected to today's practices. German counselors on the other hand, have only one eugenic memory, that of the Nazi's murderous past, which caused a strong desire in contemporary Germany to divide between yesterday's and today's practices.

Thus, while in the West and especially in Germany, "eugenics" became a word with "nasty connotations" (Koch, 2004; Paul, 1992), this is not the case in Israel (see also the section about religion and eugenics). Consequently, the present state of reproductive genetics in both societies is the theme of the next section.

MEDICAL GENETICS IN ISRAEL AND GERMANY TODAY

The Occurrence of Genetic Congenital Malformations

While the occurrence of congenital malformations in a certain society could be the reason for a wider or narrower adoption of reproductive genetics, comparing the occurrence of "birth defects" in different nations is a very difficult task, which I will not attempt due to differences in data collection and definitions of what counts as a "defect". Yet, it has been estimated that the rate of congenital malformations among the Jewish population in Israel is close to that of other Western countries (Shadmi, 2001a). This is true despite the fact that the Ashkenazi (Jews originating from Eastern Europe) gene pool is often depicted as especially prone to genetic disorders (Ivry, 2004; Remennick, 2006). Remennick (2006) does not understand the wide publicity of the "Jewish Ashekenazi diseases" to reflect pure genetic "facts". Rather, she understands it to also result from the Ashkenazim's privileged social status and involvement in the scientific world, which led to intensive genetic research into their ethnic group (see also Birenbaum-Carmeli, 2004). Subsequently, this collected body of knowledge resulted in high "genetic anxiety" among Ashkenazi couples, who erroneously interpreted the availability of tests targeting their group as a sign that they had higher overall risks than the general population

(Mishori-Dery, Shoham Vardi and Carmi, forthcoming). However, being a carrier of a recessive disorder becomes problematic only when the mating partner is taken into consideration, as consanguineous marriages dramatically raise the chances of having an affected child. Such marriages have been relatively frequent in Israel but not among Ashkenazi Jews. Rather, consanguineous marriages were common among non-Ashkenazi Jews (until lately) and among Arabs, in which this kind of mating is still very frequent and represents up to 44% of all marriages (Jaber et al., 1994), a fact which raises the general Israeli rate of "birth defects".

Thus, the warm adoption of reproductive genetics in Israel can only partly be explained by the high frequency of genetic "birth defects", especially since the population groups that use genetic knowledge and techniques most, namely the Ashkenazi couples (Remennick, 2006; Sher, 2003), are not the ones with high consanguineous mating rates and hence, with the highest frequency of "birth defects".

In the German population, there are no unusually frequent genetic diseases. Moreover, due to historical reasons, in Germany there exists no obligation to register genetically caused malformations or genetic diseases and hence, there are no accurate data available about the rate of genetic "birth defects" in Germany. Thus, the report of the German National Ethics Council (2003) explains the reduced use of screening for heterozygosity[3] in relation to certain genetic conditions in Germany, compared to Israel (as well as to Sardinia and Cyprus), to be the result of the occurrence of autosomal recessive diseases within the different populations (German National Ethics Council, 2003, p. 43).

However, in contrast to the claim of the German National Ethics Council, I do not think that these differences between Israel and Germany can be sufficiently explained by the occurrence of autosomal recessive disorders. This is mainly due to the following reasons: firstly, these differences are not "pure" genetic facts but have their own specific social history, as explained by Remennick (2006). Secondly, the Israeli population is in fact genetically more heterogeneous than the German one (and "mixed" marriages between different groups of Jews are quite common). Thirdly, it is not very clear that differences in the occurrence of congenital malformations really exist. Fourthly, some abnormal genetic conditions are quite frequent in both societies (for example cystic fibrosis or fragile X). Fifthly, interestingly, genetic screening in Israel is more common among sub-populations with a lower occurrence of congenital malformations than among sub-populations with a higher occurrence of congenital malformations. A further discussion of this issue can be found under the heading: population screening tests.

---

[3] Heterozygotes have two different alleles (any one of a series of two or more different genes that occupy the same locus on a chromosome) at one (or more) loci on homologous chromosomes. In case of autosomal recessive disorders, having only one abnormal allele means the person is perfectly healthy and thus unaware of his/her carrier status. However, if this person mates with another heterozygote carrying the same anomalous allele, their chances of having a sick child are one in four. Thus, these tests check whether the parents are carriers of the same genetic anomaly. If so, the fetus may be tested.

PUBLIC HEALTH POLICY AND THE PREVENTION OF "BIRTH DEFECTS" IN ISRAEL AND GERMANY

Health Insurance in Germany and Israel[4]

Until 1995, almost 90% of the Israeli population was covered by a comprehensive health insurance scheme that included curative and preventive out-patient as well as hospital care. Since 1995, the entire population is covered by a national health insurance, through a number of health funds. Despite the Israeli national health insurance, there is a cost-sharing amount paid for many medical services. Due to the limited coverage of the mandatory "health basket", the different health insurance companies run supplementary health insurance programs paid for privately. Likewise, private practice, mediated by social class, in which patients pay for medical care entirely out of pocket, is flourishing in Israel (Filc, 2001).

In West Germany, by 1989, before unification, the public health system covered almost the whole West German population and resembled in this regard the uniformed and centralized system of East Germany. Today, the united German health care system has one of the broadest coverages within free market societies, and amongst systems based on the social security system. With the exception of the private sector of some 8% of the German population, it does not follow a free market system (Luescen, Niemann and Apelt, 1997). There are over 1170 autonomous health funds in Germany, whose overall supervision rests with the Federal Government's Ministry of Health. Since each health fund must operate within statutory guidelines, which provide a "benefits package" that must be offered to the insured under statutory health insurance, the funds are basically similar to one another (Nippert, 1998). The funds provide full coverage for all medically necessary services such as: ambulatory and in-patient care, prescribed drugs, medical appliances, dental care, etc. The patients are free to choose their preferred physician and no money changes hands between patients and doctors.

To summarize this discussion of health insurance; whereas in Israel there is a cost-sharing amount paid for many services, this happens only rarely in Germany and overall, the German insurance is far more inclusive than the Israeli one. However, the fact that the German health insurance is both generally and specifically (regarding prenatal diagnosis, as will be elaborated later on) more inclusive, does not lead in the expected way to pushing towards more prenatal genetic tests in Germany.[5] Similarly, the lack of insurance coverage does not stop Israeli geneticists from offering tests, or well-off Israeli patients from taking them. On the contrary, it might work in the opposite direction (see the following section on cost-benefit calculations and profit motives).

[4] The data in this section concerns the time of the research and does not refer to reforms in the German health system, which occurred after the study was completed, and hence are irrelevant to our comparison.

[5] Similarly, I found no evidence pointing to the assumption that the extensive German health insurance leads to performing less prenatal genetic tests in order to save the state's money.

Prevention of "Birth Defects"

In Israel, most genetic institutes are located in public teaching hospitals affiliated with one of the four Israeli medical schools and they are organized in comprehensive, independent departments/units of different sizes within the hospitals (Chemke and Zlotogora, 1997). There are also a few private laboratories offering genetic tests. These private clinics often work together with the hospital institutes and/or with the health funds. Since 1971, the state of Israel operates a Tay-Sachs disease prevention program, which was initiated according to the World Health Organization principles, meeting its critical prerequisites (Zlotogora and Leventhal, 2000): 1. The medical condition is severe and appears predominantly in a well-defined population. 2. It is possible to detect heterozygotes by simple, accurate and inexpensive procedures. 3. It is possible to diagnose the disease *in utero*, early on in pregnancy (Wilson, 1968). In the beginning, only Ashkenazi Jews were targeted by the program (carrier frequency 1/30). Later on, the program was expanded to include Moroccan Jews, who were also found to be frequent carriers of the disease (1/60). Nowadays, as the Jewish population has become increasingly mixed, the gene frequency of Tay-Sachs disease (TSD) has increased in other communities and the current policy is to examine all individuals at reproductive age (Chemke and Zlotogora, 1997).

The Tay-Sachs screening program in Israel (which offers testing free of charge), has proven very successful. As a result, the frequency of TSD among Jews has been dramatically reduced (Kaback et al., 1993), due to selective abortions and the prevention of marriage between two carriers among the ultra-orthodox Jews. The success of this program in preventing a lethal[6] disease is part of the explanation for the "openness" of Israeli geneticists as well as pregnant women, to the adoption of genetic screening tests. β-Thlassemia is also common in Israel, mainly among the Arab population and among Jews from Kurdistan. Since carriers of this disease know of their status because it causes anemia, no general screening programs are needed. Yet, prevention programs are targeting risk populations and the tests are covered by the Israeli health insurance "benefit package", which the health funds have to follow.

Furthermore, the Israeli health Ministry also supports educational programs among the Arab population, which aim at explaining the risks of consanguineous marriages (Shadmi, 2001b) and particular genetic programs throughout Israel target specific villages and sub-populations, who are educated about the diseases common in their group and offered screening tests and PND (For a critical discussion about the Bedouins and medical genetics see Kish, 2000; Raz, 2003; Traubman, 2003a).

Since 1980, the Israeli Ministry of Health sponsors a national program for the detection and prevention of birth defects (Chemke and Zlotogora, 1997), which was

---

[6] Since Tay Sachs Disease is such a severe condition, involving mental and physical retardation as well as death at a very young age, hardly anyone argues against its prevention.

renamed "the department of community genetics in the Israeli health ministry", in the late 90s, by its than new head, Prof. Joël Zlotogora. This program includes newborn screening for phenylketonuria and hypothyroidism. It also offers screening for Tay-Sachs carriers and covers PND (mostly in the form of amniocentesis) for chromosomal aberrations in women above the age of 35 years, as well as for women at increased risk of genetic disease, which may be diagnosed prenataly.[7] In addition, the triple test is offered to every pregnant woman and is partly paid for by her health insurance (Zlotogora and Chemke, 1995). If a woman is found to be at risk as a result of the triple test and/or ultrasound screening, the health funds will cover the costs of further exams.

On the other hand, if PND is done solely on demand, the cost has to be covered by the examinee. Three ultrasounds are offered routinely to every woman during a normal pregnancy, one at each trimester. In recent years, private targeted ultrasound scans are becoming a routine check-up among pregnant women who can afford to pay for this expensive scan. Such fetal anatomic ultrasound scans (without a medical indication) are more common in Israel than anywhere else in the world (Bar and Rosin, 1998). Similarly, according to a court verdict, gynecologists in Israel are obliged to inform pregnant women about the availability of this scan, which is paid for out of pocket. Lately, the supplementary health insurance started participating in the expenses of this specialized test, as well as in amniocentesis examination for women under the age of 35.[8] In Israel, there is a national register for congenital malformations, for DS, for neural tube defects and for β-thlassemia. Access to this information is restricted.

In Germany, in 1970, PND started with the introduction of amniocentesis (AC) for the detection of genetic defects in the 2nd trimester. A registry for PND was funded by the German Research Association (DFG: *Deutsche Forschungsgemeinschaft*) for the years 1973–1979. In 1980, ultrasound (US) was introduced as a routine. Two scans were offered in the case of a normal pregnancy. Since April 1995, three scans (around the 10th, 20th and 30th weeks of pregnancy) have become the routine standard (Wegner and Becker, 1997).Yet, special fetal anatomic ultrasound scans are not commonly performed in Germany. Beginning in the early 90s the triple test (TT) has been used with increasing frequency. However, this test, which is a risk assessment test and not a diagnostic test, is not performed routinely and it has been criticized by a moratorium of the Second Consensus Meeting on Maternal Serum Screening, by human geneticists and gynecologists who advise cautious use of the test (Wegner and Becker, 1997). Ultrasound examinations and blood tests are combined in first trimester screening in weeks 11–13, which is considered to have advantages over the Triple Test, as it is done earlier in pregnancy and seems to be more reliable (National Ethics Council, 2003). Since first trimester screening does not detect the risk of neural tube defects, the "AFP test" is

[7] From 1980 to 1993, amniocentesis was offered to women over 37. In 1993 the age was reduced to 35.

[8] Information obtained in interviews as well as through telephone conversations with representatives of different Israeli health insurance companies.

also often carried out between 16 and 18 weeks of gestation. Invasive procedures (usually amniocentesis or chorionic villus sampling and more rarely, fetal blood sampling or fetal tissue biopsies) are performed predominantly when there is a higher risk than in the general population, such as a problematic family history or maternal age indication. In any case, gynecologists are obliged to inform pregnant women above the age of 34 years about the possibility of invasive techniques in PND. The costs of PND in Germany are covered by the German health insurance (through the health funds). Amniocentesis is offered on an advanced maternal basis for women above 35 years old. Yet, the health insurance funds also cover test costs for women under 35 who are afraid of bearing a child with genetic abnormalities ("maternal anxiety" indication), as long as their practitioner certifies that PND is necessary. Thus, as we have seen, German public health policy supports prenatal genetic diagnosis by all the measures mentioned above, by sponsoring studies on the development and evaluation of amniocentesis and CVS (in the past) and by the support given to hospital genetic institutes by the states (Nippert, 1992a). However, there is no equivalent program to the department of community genetics in the Israeli health ministry, (which was formerly named "The Israeli birth defects prevention program"). Such a national genetic prevention program, with its eugenicist or euphemistic title, would cause tremendous uneasiness among the German professionals and public, who repeatedly state that lowering the occurrence of "birth defects" should not be a goal of the state. Table 1 compares between Israeli and German counselor's opinions concerning this matter and it very clearly shows that whereas the majority of German counselors strongly disagree with the idea that reducing the number of deleterious genes in the population is one of their profession's goals (see also *Positionspapier der Gesellschaft fuer Humangenetik*, 1996), their Israeli counterparts are far less decisive concerning this goal and quite a few of them (up to 14.3%) even explicitly agree that cleaning the gene pool is one of their professional targets.

To sum up, both the Israeli and German states support prenatal diagnosis and reproductive genetics. Nonetheless, the Israeli state is clearly more supportive of and involved in reproductive genetics than the German state, as it operates more prevention and educational programs, especially on the community level. However,

TABLE 1. An important Goal of Genetic Counseling is to Reduce the Number of Deleterious Genes in the Population

	1993–5		2000–1	
	Israel	Germany	Israel	Germany
Strongly Agree + Agree	14.3%	1.6%	13.3%	–
Neither Agree nor Disagree	–	3.8%	13.3%	3.6%
Disagree + Strongly Disagree	85.7%	94.5%	73%	96.4%
	100%	100%	100%	100%
N	22	185	31	57

on the individual level of the pregnant woman, genetic exams (and especially amniocentesis) are more costly in Israel than in Germany. The next section will demonstrate that German women are offered far less testing than their Israeli counterparts.

Population Screening Tests

Genetic screening may be defined as any kind of a test performed for the systematic early detection of a predisposition or resistance to a genetic disease, for exclusion of such a genetic disease, or for determining whether a person carries a gene variant which may produce disease in their offspring. Such tests target low-risk populations, with no known risks (such as those having to do with the genetic history of the family or with the pregnant woman's age). Screening may be concerned with the general population or with specific sub-populations defined on some basis other than their health. Genetic screening is distinguished from other types of medical screening by the genetic nature of the disorder, which may result in risk implications to family members of the person screened, even though they might not wish to be included in the screening program. Its aim is also not necessarily to prevent or treat disease in the person screened but rather in his/her offspring.

The most dramatic difference between the two fields of reproductive genetics here under study can be found in their practice of heterozygotes population screening tests. Such tests are very common in Israel, as they are widely publicized, offered by all genetic institutes and partly covered by the supplementary health insurance. During the 90s in Israel, different heterozygote screening tests (especially targeting non high-risk pregnant women, namely women without a problematic family history) started being offered on a partly private basis by genetic hospital units, as well as by a few private laboratories. According to the interview with Joel Zlotogora (Head of the department of community genetics in the Israeli health ministry), population screening tests on such a wide scale are not known to exist outside Israel, despite the fact that some relatively severe genetic anomalies that are tested for in Israel (such as CF and fragile X syndrome) occur everywhere. Recently, the Israeli Association of Medical Geneticists in its position paper called: "Screening for the detection of carriers of genetic diseases" (2004), advised different sub-populations (depending on occurrence rates) to check for the following conditions: Tay-Sachs, β-thlassemia, cystic fibrosis, familial dysautonomia, Canavan Disease, Costeff syndrome, metachromatic leukodystrophy and fragile-X (which is offered to all groups). The Association less strongly recommended testing for Fanaconi anemia (type C), Bloom syndrome, Niemann-Pick Type A, mucolipidosis IV and ataxia-telangiectasia.

Other tests are offered in hospitals and labs across the country but have been declared controversial by the Association of Medical Geneticists, such as tests for Usher syndrome 1 F, alpha 1-antitripsin deficiency (allele Z), GSD1 and limb dirdle muscular dustrophy 2B (dysferlin). Other tests such as Gaucher, familial Mediterranean fever, albinism, non-syndromic deafness, phenylketonuria, maple syrup urine disease and alpha 1-antitripsin deficiency (allele S), were not recommended by the association (Erez, 2004). And yet, they are also preformed.

Of this entire set of tests, Ashkenazi couples may be offered up to 14 tests, while non-Ashkenazim are usually offered 2–3 tests. If one of the mating partners is found to be a carrier, her/his partner is also examined and if both are carriers, it is usually advised to test the fetus. The first round of tests for any of the partners, is not covered by the department of community genetics in the Israeli health ministry or by the "benefits package" of the health ministry.

According to Zlotogora and Leventhal (2000), these private screening programs have led to considerable confusion, not only for health professionals but also for the public, which lacks information. In most hospitals, the private screening programs are offered with very little pre-testing explanations (usually the women are given a sheet of paper with a few words on each disease) and with no pre-testing counseling. The recent inclusion of some of the tests in the supplementary health insurance provided by the health funds (for which extra money is paid by the insured), added to the confusion. The reasons for not funding these more recent screening tests in Israel are not ethical (like in Germany, where heterozygote screening tests are strongly opposed) but financial. One group that has a vested interests in maintaining the *status quo* is the health funds themselves, who by partly covering these tests, which are popular among the better off and better educated, young segments of the Israeli population, encourage their members to pay for additional insurance. Other parties that enjoy the financial fruits of the tests are the genetic hospital units themselves. Yet, for them it does not matter whether funding comes out of patients' pockets or out of state or health-fund budgets. Subsidizing such tests could potentially increase their number but, on the other hand, if the health ministry will work only with one laboratory offering the best price (like in the case of Tay-Sachs), hospital institutes may not benefit. In 2000, the Israeli Association of Medical Geneticists advised the health ministry to include some of the discussed screening tests (for fragile-X, CF, Canavan disease, and a few other rare conditions found among specific populations) in the health "benefit package" (An Offer for a New Technology, 2000). The association, referred to cost-benefit analysis, and mentioned the potential economic benefits of such tests for the state, but the offer has been rejected up to now.

In Germany, there is a professional reluctance to be associated with population screening tests (Harris and Reid, 1997), as similar and even higher carrier frequencies among the German population for genetic anomalies are not considered by German counselors as justifying population screening. Consequently, population screening tests are not offered to the German public (Harris and Reid, 1997). Argumentation concerning cystic fibrosis (carrier frequency of about 1/30), which is the most common life-shortening autosomal recessive disorder in Germany, will serve to demonstrate the logic behind this fact. CF affects the respiratory, gastric intestinal and reproductive organs and the sweat glands. The median age at death due to CF has increased over the years, yet it is only 31.6 years (CFF Patient Registry, 2002). Irmgard Nippert (1998) has interviewed health care providers in

Germany about CF screening and found out that it is generally agreed that CF screening in Germany should only be offered on a strictly individual basis and should not become a standard of care in pregnancy (like in Israel). The majority of health care providers interviewed in Nippert's study, thought that active screening programs harbor the potential of eugenics and should not be pursued. Similarly, the German CF Association opposes CF screening on ethical grounds (see Chapter 7). Moreover, whenever the reasons to oppose screening in Germany are not purely ethical, they are bureaucratic. For example, the German society for human genetics, which is not a very strong opponent of genetic screening, acknowledged in its guidelines from 1996 (*Positionspapier der Gesellschaft fuer Humangenetik*, 1996) that the development of new tests is currently growing and will detect a large number of diseases, which could be screened on a population basis. Yet, the society supports that tests should be available only for informed counselees, i.e., for people who have knowledge about the disease, for example family members of an effected individual, but not on a wider basis. The reasons given for the rejection of general screening are bureaucratic, namely that at that point in time the public was not well enough informed and there were not enough counselors available to do the counseling for every single pregnancy. One can obviously question whether this is a good enough reason to deny future parents the information and tests that could help them to avoid having a sick child. In a non-optimal situation where thorough counseling is not available, for instance, for every CF screening test, the question obviously becomes: which situation should be avoided? One in which because parents were not trusted to handle the test in their own best interests without counseling, they may have a child with CF, who's birth they never had the chance to prevent. Or a situation in which many parents would have more knowledge and autonomy, which would go hand-in-hand with unnecessary anxiety and sometimes with the need to make very complicated decisions quickly, without proper support beforehand. In Israel, as opposed to Germany, the second option is preferred, as CF tests are widely offered. This obviously has to do not only with bureaucratic reasoning but also with the lack of ethical opposition to selective abortions of embryos with CF, as will be discussed later on (see Chapters 5 and 7).

To sum up, not only the state, but even more so the medical establishment in Israel is pushing towards a wide use of screening tests, as opposed to the German medical establishment, which is responsible for not introducing such tests to the general public. However, the German public itself is not washing its hands of prenatal genetic tests, as the next section will show.

THE PUBLIC'S UTILIZATION OF TESTS AND SELECTIVE ABORTION PRACTICES

In the previous sections I have discussed the genetic prenatal testing possibilities opened up by the Israeli and German states and health establishments. Let me now take a look at the public's side of things. Due to the structure of this study, I cannot attest with my own data to German and Israeli women's knowledge about reproductive genetics or to their logics of justification concerning reproductive decisions. However, other studies have shown that large gaps exist between different groups of women within the same nation, between the public's and expert's opinions concerning the uses of prenatal diagnosis and between public policy and the public's opinions and uptake of tests. In Israel, it has been shown that the documented boom in reproductive genetics is not similarly experienced by all women. In fact, social inequality and cultural differences are apparent in women's practices, as the utilization of tests and the decision to carry out selective abortions differ among diverse population groups. Being Jewish, secular, educated, having a higher income and a private health insurance, as well as having fewer children, being of Ashkenazi origin, being over 35 years of age and having knowledge with regard to prenatal diagnosis, were all found to be significant factors in predicting the performance of genetic tests in Israel (Mishori-Dery, Shoham Vardi and Carmi, forthcoming; Sher et al., 2003). Sher et al. (2003) have also shown that 60.9% of Jewish Israeli women take the "triple test", 63.3% take the Tay-Sachs carrier test, 24.3% take part in fragile-X carrier testing, and 50.8% of the women aged 35 and up have amniocentesis tests. In Germany, According to Nippert (1992a), it was estimated that about 50% of all women eligible for prenatal genetic diagnosis on the indication of the age of the mother-to-be, obtain it, whereas the report of the German National Ethics Council (German National Ethics Council, 2003) estimates that only about 20% of women at increased risk in Germany refuse invasive prenatal diagnosis. The women who obtain the test tend to be middle and upper-middle class, highly educated and living in urban and suburban areas (Nippert, 1992a). Thus, it seems that the women's compliance to genetic tests with which they are familiar, namely amniocentesis, is not dramatically different in Israel and Germany (and may even be higher in Germany), as in both countries the middle and upper-middle classes, as well as the more urban and educated, are more interested in being tested. Thus, the Israeli counselors' claim that they offer many tests simply because they are attentive to the public's demand, which has been repeatedly mentioned in the interviews I conducted in Israel, seems problematic. In fact, the German case teaches us that without the support of the medical establishment and the state, it is very difficult for the public to develop its "own" demands. Additionally, Mishori-Dery, Shoham Vardi and Carmi (forthcoming) have demonstrated that health professionals in Israel are significantly more supportive of comprehensive prenatal testing than women, the potential consumers of the tests. In contrast, studies about the related topic of preimplantation genetic diagnosis (PGD), which is currently prohibited in Germany, have shown the German public to be more in favor of allowing PGD than certain expert groups (mostly midwives and ethicists) and the current policy (Krones, 2005; Krones and Richter, 2004).

Returning to the women's practices, Erikson (2003) has argued that when faced with a "positive" diagnosis, most German women opt for selective abortion. Likewise, she argues that German doctors clearly favor selective abortions in case of a discovered "pathology", in an unspoken but ever-present manner. Similarly, the report of the German National Ethics Council (2003), states that the number of invasive prenatal examinations in the Federal Republic of Germany is constantly increasing. However, if this is really so, it still remains to be explained how is it that while 17% of all legal abortions done in Israel in 2003 resulted from an embryopathic indication (www.health.gov.il), in Germany the estimated[9] rate of post-diagnostic abortions is only 2–4% of all recorded terminations (National Ethics Council, 2003). Similarly, late abortions (after viability), performed mostly for embryopathic reasons, are far more common in Israel than in countries such as Germany (see Chapter 5), the US, the UK, Denmark and Canada (Gross, 1999). Based on my empirical study, no conclusive explanation can be given for these dramatic differences, as I did not study the actual practice of selective abortions in both countries. Yet, part of the answer clearly lies in the fact that the power social policy and the medical professionals' hold restricts the possible uses of medical genetics in Germany, as it forms a barrier between what is desired by patients and what is known and offered to them. In contrast, the encouragement of genetic tests by the Israeli state and medical establishment leads to more selective abortions, which may also be due to more detected anomalities, resulting from a larger performance of tests. It is hard for me to tell whether it is due to the discussed different policies that more anomalies are discovered in Israel than in Germany and thus more pregnancies are terminated. However, throughout this work I will argue that different perceptions of what counts as a life (un)worthy of living in both societies, contributes both to a differential performance of tests and to a differential understanding of what kinds of "positive diagnosis" are severe enough to justify an abortion, by both women and professionals.

IS THERE A REDUCTION IN GENETIC "BIRTH DEFECTS"?

Despite clear efforts to lower the rate of birth defects in the Israeli population, the success of such efforts is unclear. Generally, there are no exact data available in Israel about reduction in birth defects and the existing data have to be read carefully, taking into consideration the lack of complete registration. For example, Tay Sachs disease has almost disappeared but other genetic problems have not. Otherwise, the prevalence of Down's syndrome for example, decreased in the late 80s and early 90s, probably due to acceptance of AC and selective abortions (Health Condition in Israel, 1999), but it increased again in 1997–2000. Likewise, in 1995, 1.62% of Israeli-born babies (out a total of 116,886 for that year) were born with birth defects, which have to be reported to and registered by the health ministry. In 1999 the percentage climbed to 1.91,[10] and between 2000 and 2003 a 22% increase in the number of registered "birth defects" occurred (Trabelsi-Hadad, 2005). Likewise, consanguineous marriages among the Arab population remain very common, despite

[9] The numbers in Germany can only be estimated, since the 1995 German abortion law has abolished explicit embryopathic indications and thus, such abortions are not registered. See Chapter 5

[10] Data obtained from the Ministry of Health, public health services, mother, child and adolescent health department, Jerusalem.

different educational efforts to lower their occurrence. In Germany, the impact of PND on the prevalence of genetic disorders can be shown most reliably for trisomy 21 (Down's syndrome). Studies clearly demonstrate an increase in the prenatal detection of trisomy 21 (Sperling et al., 1994), and almost all of these pregnancies are terminated. Otherwise, there is no data for the effectiveness of genetic services in Germany (Nippert, Horst and Schmidke, 1997).

PUBLIC OPPOSITION TO REPRODUCTIVE GENETICS

Reproductive genetics has encountered very different kinds and levels of public opposition in Germany and Israel. Answering the question "Do any of the following act as sources of opposition to genetics services in your country"? The geneticists from the 1993–95 Wertz and Fletcher international survey answered as follows (see Table 2):

Table 2 presents the fact that reproductive genetics faces very little opposition in Israeli society, most of it coming from religious groups and not from women's groups, disability organizations or political parties (since the opposing Israeli parties, are the religious parties). On the other hand, unusual coalitions formed between religious parties, which allied with the green party, feminists and disability activists in Germany, have hindered the adoption of reproductive genetics in Germany (Krones, 2005; Nippert, 1992a). These groups are unified in their fight for the protection of what they understand as the value of life with disability, the protection of human dignity and rights of the embryo, and the struggle against the revival of eugenics, a combination which makes the German ethical debate fundamentally harsh (Krones and Richter, 2004). Thus, public ethical debate in addition to legal debate concerning reproductive technologies is flourishing in Germany (Lanzerath, 2004), as opposed to Israel, where it practically does not exist. This difference can be best demonstrated by pointing to the different commissions that have been appointed by the German parliament as well as by the German Chancellor during the last 20 years to work on recommendations for the regulation of PND and PGD techniques (Bericht der Enquete-Komission, 1987, "Chancen und Risiken der Gentechnologie", German Bundestag, 2002, Study Commission on "Law and Ethics in Modern Medicine" and National Ethics Council, 2003, "Genetic diagnosis

TABLE 2. Opposition to Genetic Services

	Israel (% checked)	Germany (% checked)
Religious groups	68.2%	56.2%
Women's groups	4.5%	61.8%
Political parties*	18.2%	28.3%
Advocates for Persons with disabilities	4.5%	74.1%

* *In Israel, the opposing political parties are religious parties.*

before and during pregnancy"). No similar committees have been appointed in Israel. Whilst this entire study looks into the specifics of the German and Israeli debate or lack thereof concerning PND, and into the precise laws regulating this field, in the following section I wish to elaborate on the specific influence of religion on the bioethical discourse in both countries, an influence which is often overlooked, especially in scholarly discussions of German bioethics. I suggest that the comparison with Israel, one of the few Western countries in which bioethics discourses lack Christian embeddings (Prainsack, 2006a), makes this suppressed influence far more salient.

RELIGIOUS THOUGHT, EUGENICS, AND "PLAYING GOD"

Adopting new genetic reproductive technologies provokes moral dilemmas related to the issues of abortion, human dignity and rights of the embryo and the disabled, eugenics and interference with nature or, in religious terms, of "playing God", all of which have moral roots in religious doctrines. While chapter five thoroughly discusses the status of embryos according to Christian and Jewish teachings and chapter seven discusses religious thought concerning the disabled and human suffering, in this section I will focus on religious attitudes towards eugenics and towards the idea of "Playing God". However, my claim is not that concrete religious practices and beliefs shape the field of reproductive genetics. Rather, religious ideas function as moral-mental structures, which linger on in contemporary society (Latour, 1993; Rabinow, 1999; Shenhav, 2005) and can thus be traced in the way modern scientists, policy makers, ethicists as well as the disabled, interpret the moral dilemmas associated with reproductive genetics. The relationship between traditional religious ideas and modern scientific innovations and techniques is a complex one and does not always work in a foreseeable manner, which restricts scientific innovations and their uses. However, returning to the findings presented in Table 2, it seems as if in both Israel and Germany, religion serves as a barrier for the implementation of repro-genetic technologies. Furthermore, in the interviews I conducted with genetic counselors in Israel, a conflict between (mostly secular) counselors and religious counselees and authorities was often mentioned. Many Israeli counselors expressed a lot of resentment towards religious patients and their Rabbis', basically complaining that the Rabbis' anti-abortion verdicts turn into law suits against themselves and their hospitals. Furthermore, most counselors thought religious people were backward or "primitive"; one Israeli counselor even explicitly argued that if the religious knowingly choose to have "sick" children, he did not think "people like us" (he and I) should pay taxes to support those "miserable children". The same conflict was also described in 1989 by Chemke and Steinberg (1989), two prominent Israeli geneticists, who wrote:

"A good number of clients of genetics in Israel are religious and obey the dictums of the rabbis in accordance with Jewish law. The geneticists themselves, however, are mostly irreligious, following the Western secular philosophy as their personal approach to the medical dilemmas in medical genetics. This creates some tension in certain communities: on the one hand, disagreement with the counseling,

particularly when it results in abortion, causes lack of confidence in physicians and suspicion of misconduct by the medical community on the part of the rabbis and their followers. On the other hand, there is anger and occasionally hatred on the part of the medical community at large and geneticists in particular, because of their feeling that religious groups do not understand the purposes and importance of genetic services"

Yet, I argue that things are not exactly as they seem. Rather, I maintain that despite the discussed resentment and since every social-moral system draws upon religious values prevalent in the history of that society, Israeli "secular" counselors are much more "Jewish" in their way of reasoning, than they themselves imagine. Likewise, Jewish religion is more supportive of reproductive genetics than it may seem at first glance or in a non-comparative context. Thus, in the following I will argue that religious Jews and genetic counselors in Israel seem to share the same implicit logic concerning PND, despite their explicit disagreements. Consequently, while it is true that most Orthodox Rabbis and their followers oppose selective abortions, they **do not** oppose preventing affected pregnancies or the life of disabled people, a factor that is one of the main reasons behind the warm endorsement of reproductive genetics in Israel.

By contrast, in Germany criticism of reproductive genetics is widespread and the churches are clearly on the side of the opponents to many reproductive technologies. However, while in Israel resentment of counselors is focused at religious authorities, my interviews in Germany revealed that German counselors perceive themselves to be in conflict with political parties (such as the Green party) and disability activists, more than with religious authorities. Yet, as in the case of their Israeli counterparts, I will argue that German professionals as well as the German ethical and legal discourse, lean on religious cultural understandings while trying to handle the dilemmas put forward by the controversial practice of reproductive genetics and that the Christian religion is an important reason behind the cautious endorsement of reproductive genetics in Germany. Moreover, the fact that religious authorities ally with the Green party as well as with the disabled and the feminists in Germany, in somewhat surprising coalitions (Krones and Richter, 2004; Schlegel, 1997), further explains the cautious adoption of reproductive genetics in Germany.

While not denying the strong effect of the Nazi recent past on the differential wish of German and Israeli experts and policy makers to detach from eugenics and its immoral connotations, in this section I focus on how religious tradition is related to eugenics and thus, to contemporary practices and their moral acceptance. Concerning Judaism, Immanuel Jakobovits (1967) contends that Jewish law has always been supportive of eugenic ideas (even before modern genetics was founded), as it encouraged individuals to be responsible to society and to the generations yet unborn in a manner that was unknown in any preceding system of religious thought or social medicine. In addition, different provisions in Jewish law from medieval times were clearly motivated by eugenic considerations for the moral excellence of the progeny, which they associated with physical excellence, as they prohibited marriage into families with hereditary disorders (see also Feldman, 1998). By contrast, in his essay about Jewish eugenics, Noam Zohar (1998) contends that the conventional views which perceive Jewish tradition to be eugenic and one may even say racist (with its idea of the "chosen people"), involve a one-sided reading of the Jewish tradition and particularly, the suppression of traditional critiques of lineage and of the notion of a "Jewish race". Likewise, Barilan (forthcoming, a) also disagrees with Jakobowitz on this point. Without going into this theological argument, it is important to draw the links between Jewish attitudes towards the health of offspring, and contemporary Jewish Orthodox practices. The present Orthodox Jewish community is utilizing contemporary genetic knowledge in a worldwide unique fashion, as it runs a program called "*Dor Yeshorim*", for carrier screening tests[11] prior to prearranged marriages, which is the common way to marry in this community (Prainsack and Siegal, 2006). The purpose of the program is to prevent the mating of two heterozygotes and thus to prevent affected pregnancies and the life of disabled people (prior to conception). Screening is strictly anonymous and results are provided to the matching organization. If both potential partners are found to be carriers of the same abnormal allele, the process towards engagement stops. Prainsack and Siegal (2006) claim that the fact that "*Dor Yeshorim*" gives no information on individual disease carrier status but only on the "genetic compatibility" of prospective partners, allows the avoidance of pressing issues with which "secular" genetic screening programs struggle, such as passing on too much "useless" information to the tested individual. In my opinion, it is also demonstrative of the fact that Jewish culture is in favor of preventing life with disability. Moreover, this *prima facie* positive assessment of reproductive genetics has also been found to characterize Israeli disability activists. In interviews with chairpersons of Israeli disability organizations, Raz (2004) has found that they perceive prenatal genetic testing to be eugenic. However, they supported it for precisely that reason, as eugenics has no negative connotations for them, implying only an improvement in the health of their progeny (see also Chapter 7).

In contrast to Judaism, according to Bassett (1995) the only Christian laws that have any eugenic significance are those prohibiting consanguineous marriages. More so, the traditional view of Christianity and particularly that of Catholicism, rejects a eugenic world view (Inge, 1968). Even in Germany between 1920 to 1945, although some forms of Catholic eugenics existed, by and large, the Catholic church opposed eugenics as well as abortions and sterilization, on the grounds of the threat to the integrity of the human individual, which such ideology and practices pose. As Dietrich writes (1992): "The Catholic hierarchy and theologians had generally opposed the compulsory negative eugenics policies of the Nazis and their predecessors, interpreting them as intrusions into the private spheres of human relations and as an unnatural interference with God's creative plans…" (p. 576). In fact, in 1930 with the promulgation of the Papal Encyclical, *Casti Connubii,* compulsory sterilization for eugenic reasons was forbidden by the church.

---

[11] The current list of screened diseases includes: Tay-Sachs, cystic fibrosis, Gaucher's disease type I, Canavan disease, familial dysautonomia, Bloom syndrome, Fanconi anemia, glycogen storage disease type 1A, mucolipidosis type IV and Niemann-Pick disease type A. (See Prainsack and Siegal, 2006).

An example worth mentioning in this context is Cardinal Galen, who in 1941 endangered himself by openly condemning the Nazi euthanasia program; Galen remained, however, silent about the fate of the Jews.

According to Walter (1999), theological moral judgments concerning modern genetics are shaped by theological models of *Imago Dei* (human beings having been created in the image of God) and the tendency of humans to "play God". These are also different in Christianity and Judaism. According to Barilan (forthcoming, a) rabbis never ban issues such as stem cell research, cloning or genetic experimentation on humans *a priori*. Rather, they try to find legal solutions, which enable them not only to tolerate such practices but also regard them as morally valuable. This is in line with the fact that the *Halakhah* usually sees its role as lifting religious and other impediments on medical practice and progress and not as its watchdog (Barilan, forthcoming a). Furthermore, while most Christian teachings consider humans to be subordinate to God in the process of creation, in Judaism there can be no accusation of "playing God" as human beings are understood to take an active part in God's creation, which they are obliged to improve. Human dignity is thus understood to manifest itself in the willingness to alleviate suffering and in this way, to improve on God's creation (Heyd, 1992; Prainsack, 2006a; Wahrman, 2002). While a similar view to the Jewish one, which understands humans to have a responsibility to help bring creation to completion, exists in Christian doctrine, another dominant perspective understands intervening in the very material that constitutes life as "playing God", and thus as exceeding human limits. This understanding emanates from the stewardship model, which tends to limit human activity, as it stresses preserving creation (Walter, 1999). To conclude, in the Jewish world reproductive genetics is adapted to contemporary Jewish Orthodox life, and can be associated with science's blessings and thus, accepted as an unquestionable medical routine. In the Christian world, on the other hand, it is harder for it to be accepted due to the absence of any religious eugenic heritage. Furthermore, in a Christian context, the fear of the harms associated with science's transgression into God's creative plans is more prevalent.

FEMINISTS

Other oppositional groups mentioned by German geneticists (see Table 2) are women's groups and advocates of people with disabilities. While chapter seven discusses the positions of people with disabilities and their advocates towards reproductive genetics, I would like to address here the positions of feminists in both societies towards this practice. Reproductive genetics and selective abortions have posed ethical conflicts for the feminist movement, which traditionally supported abortion on demand as one of its main platforms. This is since the abortion of a future, initially wanted child due to its genetic traits, has been said to resemble discriminatory practices directed against women on the grounds of their genetic make-up, namely their sex. Thus, the almost automatic position of feminists in support of abortions has been complicated by selective abortions

(Asch and Geller, 1996). However, in Israel no feminists have argued against prenatal genetic diagnosis or selective abortions. Whereas some writers such as Weiss (2002), Ivry (2004), Remennick (2006), as well as myself, have pointed at the burdens reproductive genetics puts on women, none have argued strongly against this practice. In contrast, German feminists often understand new reproductive technologies to be an attempt by patriarchy to oppress women, a claim promoted especially by feminists coming from the former West Germany (Erikson, 2003). In fact, some German feminists have argued that selective abortions are wrong under any circumstances and that reproductive genetics deprives women of their innate feminine power (Erikson, 2003). Likewise, unique in the international feminist landscape, German feminists often do not understand abortion to be a woman's right to refuse to have an unwanted child but rather a situation comparable to killing someone else (namely the embryo) in self-defense (Krones and Richter, 2004). Similarly, of the many feminist organizations in Germany that have expressed a view on preimplantation genetic diagnosis (PGD), none has spoken in favor of this controversial technique (National Ethics Council, 2003), despite some disagreements among themselves. Moreover, women who choose genetic testing during pregnancy are accused by German feminists of not having the right moral consciousness, or being influenced by rigorous concepts of normality (Krones et al., 2006). Once again, different cultural logics seem to facilitate different positions also among women's groups, as while Israeli feminists did mention the possible abuses of new reproductive technologies, the general German position has been far more unsympathetic towards them.

PROFESSIONAL GUIDELINES AND ETHICAL DISCUSSION

Whereas in Israel there is no wide and continuous discussion regarding the ethical issues involved in medical genetics, Germany is characterized by a profusion of such discussions. Consequently, Israeli genetic counselors have not issued their own guidelines. According to Joël Zlotogora, head of the department of community genetics in the Israeli health ministry, Israeli counselors follow the American guidelines, as they support patients' autonomy, informed consent and non-directive counseling, which are the standard ethical principles of Western prenatal genetic diagnosis. Yet, medical geneticists in Israel have often complained about the lack of clear and more specific guidelines, a situation which leads to unnecessary tests and to the commercialization of this medical field (Zilber, 1999). Writing about ethics and medical genetics in Israel, Chemke and Steinberg (1989) mentioned that ethical discussion, ethical codes and professional guidelines for human geneticists do not exist. Rather surprisingly, this has not dramatically changed in recent years. The more so, even when position papers concerning the performance of different screening tests have been issued by the Israeli Association of Medical Geneticists, they have not necessarily been followed by all counselors.

A good example of such a situation is the history of Gaucher disease screening. Gaucher disease is an inherited deficiency of the glucosidase enzyme, resulting in the buildup of a toxic substance (glucosylceramide) in different parts of the body, such as the spleen, liver and bones. Non-neurophatic Gaucher disease is frequent in the Ashkenazi Jewish community. Its most frequent mutation, N370S, is often associated with a mild type of Gaucher disease, while the second most frequent mutation, 84GG, is associated with a more severe disease. The frequency of the mild form of the disease, the inability to predict the severity of the phenotype and the existence of an effective treatment for this phenotype, were among the arguments against screening for carriers of Gaucher disease in the Ashkenazi Jewish population (Zlotogora and Chemke, 1995). Yet, despite clear voices calling for a halt to this test, most genetic units in Israeli hospitals have offered and are still offering them.[12] Likewise, screening for hearing disabilities has lately been opposed by the association of medical geneticists and yet, such tests are still performed in some genetic clinics. This lack of explicit ethical discussion, written policy and guidelines, as well as uniformity among different hospitals, so characteristic of decisions concerning the beginning of life in Israel, has also been found to characterize end-of-life decisions in Israel (Sturman, 2003). While repro-genetics is not regulated, the Israeli Law on Genetic Information (2000) does regulate medical genetics by demanding informed consent for collecting DNA samples, regulating the testing of minors, sheltered and the legally incompetent and by regulating the passing on of genetic information to a patient's relatives. This is allowed even if the patient disagrees, after getting the permission of an ethics council, which is convinced of the necessity for such an act.

In sharp contrast, Germany is characterized by a proliferation of professional-ethical discussion, as can be demonstrated by the fact that three different German professional groups have issued guidelines concerning PND: the German Society of Human Genetics (*Positionspapier der Gesellschaft fuer Humangenetik e.V.*, 1996), the Federal Board of Physicians (*Leitlinien zur Erbringung humangenetischer Leistungen: 1. Leitlinien zur Genetischen Beratung, Berufsverband Medizinische Genetik e.V.*, 1996) and the German Physicians Professional Association (*Richtlinien zur pränatalen Diagnostik von Krankheiten und Krankheitsdispositionen, Bekanntmachungen der Bundesärztekammer*, 1998). Generally speaking, all the German guidelines repeatedly emphasize non-directive counseling and the autonomy of the individual decision making of patients. More specifically, all guidelines oppose active counseling, namely, reaching out to other people, mostly family members, without their direct wish. This is since according to the autonomy principle, it should be open to the consultant to tell family members about the findings but the counselor is not allowed to give any information to a third party. Furthermore, all guidelines stress freedom of reproductive choice, which also means securing the freedom to carry an affected fetus to term. Therefore, they stress that PND should be available to all women, even when they oppose abortion. In this

[12] For more information about Gauche disease and Public Policy in Israel, see Gross, Michael, 2002, and Traubman, Tamara, 2003.

case, PND is meant to give parents more time to prepare for living with a disabled child. Moreover, in all the guidelines, genetic counseling prior to as well as after genetic testing (in case of positive results) is regarded as indispensable and respect for the individual patient's moral values or religious beliefs, as well as psychosocial situation is demanded. While the parent's wish for a healthy child is respected by all guidelines, it is repeatedly mentioned that this position can bring about a conflict with the protection of the unborn and a legal conflict between the fetus' right to life and the mother's/parents' personal right to freedom of choice, as both these positions are protected by the German Basic Law. Such a conflict is not discussed in Israel, since the Israeli legal system does not generally protect fetuses, (see chapters 5 and 6). Furthermore, gene pool cleansing aims are strongly opposed by all guidelines. For example, the society for human genetics explicitly disassociates itself from such goals, as its guidelines forbid reducing the prevalence of specific non-curable diseases or disabilities within the population, or subgroups within the population, if this reduction can only be achieved by influencing single individuals in their process of decision making. That is because according to the society, influencing an individual to take steps in order to better her/his group's gene pool would harm her/his dignity. Yet, the guidelines acknowledge the fact that the decrease of prevalence of a genetic disease could be a side effect but never the main aim, of applied human genetics. What is more, the guidelines of the Society for Human Genetics are written in a self-reflective way, as they acknowledge medicine's lack of scientific basis for defining what is "normal", or for deciding about the aims of improvement. Furthermore, they state that the present genetic variability, including all natural extremes of manifestation, should be regarded as the normal characteristics of a population. German guidelines are in line with most Western European countries and the USA, and they are also partly in line with recommendations by the Human Genome Organization (HUGO) and the World Health Organization (WHO) (Nippert and Wertz, 2001). Having described the basic principles of the different guidelines, it remains to be asked how effective are they? According to Nippert and Wertz (2001), this is not clear. Research has shown that guidelines that were developed by a specific professional group are usually ignored by other professional groups. This happens, for example, when gynecologists perform AC or CVS without prior counseling, which is against the principles of the counselor's guidelines (Nippert and Wertz, 2001).Yet, my impression from the interviews is that German counselors are aware of their profession's guidelines, and albeit sometimes critical of parts of the ideas embodied in the guidelines, they do serve as common ground for forming their ethical positions.

COST-BENEFIT CALCULATIONS AND PROFIT MOTIVES IN ISRAEL AND GERMANY

Cost-benefit analysis related to genetic abnormalities is not ethically rejected in Israel. This can be demonstrated by the following studies: a cost-benefit analysis of prenatal diagnosis and early detection of genetic disease was carried out in Israel in the fiscal year of 1985/6 within the framework of the National Program for the Prevention of Birth Defects. It was estimated that for each 1 US$ spent, 55 US$ were saved (Chemke and Zlotogora, 1997), that is, since the interruption of pregnancy is cheaper than hospitalization, medication and other health care costs of children with genetic abnormalities. Similarly, in 1994 Ginsberg et al. wrote in favor of CF screening after performing a cost-benefit analysis in Israel (Ginsberg, 1994). In a similar manner, in 2000 the Israeli Association of Medical Geneticists advised the health ministry to include certain screening tests in the "health basket" (An Offer for a New Technology, 2000), mentioning the potential economic benefits for the state of such tests. On the other hand, in Germany, the difficult ethical problems associated with PND have made public health administrators, as well as physicians and politicians, especially careful about economic analysis of PND. No-one wants to be accused of preventing the life of handicapped children for economic reasons and thus decisions about selective abortions are commonly presented as matters for individuals and the problem is constantly relocated to individual decision making (Nippert, 1992b). Writing a chapter about Ethics and Medical Genetics in the FRG in 1989, two prominent German human geneticists, T.M. Schroeder-Kurth and J. Huebner, wrote about cost-benefit analysis:

"A short comment will suffice: it is not our purpose to add anything to the calculations that have been done by others... it is ethically objectionable to describe the life of the handicapped human being in economic terms per year, although such calculations must be made by politicians to coordinate social and medical programs. ... However, we strongly oppose a cost-benefit approach to the lives of handicapped persons. .."[13] *(Schroeder-Kurth, and J Huebener, 1989. p. 163).*

This does not mean such calculations are not done in the real practice of genetic counselors. In the interviews I conducted in Germany, I heard stories about cases in which the counselor had asked the health fund to pay for an expensive PND test. The positive answer given by the health fund was justified on the basis of the costs of raising such a child, in case the fetus was found to carry the disease and was not aborted. Yet, counselors said such a justification is usually not stated openly or written down. Some of my interviewees described this phenomenon as "typical German hypocrisy".

Thus, the discussion of cost-benefit calculations in both societies shows that in Israel economic discourse is replacing moral philosophy and substantive rationality, which is considered with ethical argumentation, is being instrumentally defined by professionals and thus replaced by formal rationality (Weber, 1968), essentially without any hesitation.[14] In contrast, in Germany this cannot be explicitly done, as criticism of the instrumental logic of modern society and especially of its scientific

[13] Schroeder-Kurth, T. M. and Huebener, J. 1989. "Ethics and Medical Genetics in the Federal Republic of Germany", pp. 156–175 in *Ethics and Human Genetics: A Cross-cultural Perspective*, edited by Wertz, D.C and Fletcher, J.C. Berlin: Springer Verlag.

[14] For an analysis of a professional discourse in terms of instrumental rationality see Shenhav, 1995, 1999.

institutions, is far more embedded in public as well as in professional discourses, (as will be further elaborated under the heading: Scientific Mentalité).

Another economic aspect of PND clearly has to do with professional profit motives, which are differently organized in Israel and Germany. Historically, genetic counseling in Germany has been located in public settings, a situation which is unique in the overall German medical system. After WWII, Germany's Health insurance model, dating to the days of Bismarck, was restored and in this process, the medical profession won a dominant role for solo office-based physicians for ambulatory care, prohibiting industrial or public health doctors from treating patients. It is since then that West German physicians work entirely in private practice or entirely in a hospital, as the dichotomy between ambulatory and in-patient practice is statutory and strictly enforced. Consequently, most hospitals are not allowed to operate out-patient departments, as they may intrude on the private practitioners' monopoly only if they are affiliated with a medical school and hence, their out-patient clinic serves a teaching function (Nippert, 1992a). In the case of human genetics, the basic rules of the German health care provision were bent, as genetic counseling and PND, which are basically out-patient services, were historically only offered by hospitals affiliated to medical schools. In the early 70s, when genetic services took off in Germany, the reason for this was the scarcity of the special skills required for genetic counseling and the lack of possible profit for practitioners working in this field (Nippert, 1998). Today, this structure is being changed, as all procedures for invasive and non-invasive PND are done in public as well as in private settings. In fact in 1995, more than 50% of the German facilities specializing in genetic counseling and diagnosis were in private practice (Nippert, Horst and Schmidke, 1997) and in 1997 about 40% of the Cytogenetic and DNA laboratories in Germany were run privately (Wegner and Becker, 1997).

In contrast, Israeli genetic counselors conduct most of their work in public settings (see Chapter 4), although the superior among them may also operate private clinics. However, a public setting does not necessarily mean patients do not pay out of their own pockets. As a matter of fact, although most Israeli counselors work in public settings, (as opposed to most German counselors, who work in the private sector), more Israeli patients pay for genetic services with their own funds, as genetic hospital units in Israel charge patients directly (especially for screening tests).

On the contrary, while most German counselors work in private clinics, most tests are paid for by the comprehensive German health insurance, as there are no economic hindrances to genetic service access in Germany (Nippert and Wolff, 2004).

Table 3 describes the differences between Israeli and German counselors' answers to the question: What percentage of your patients have to pay for genetic services from their <u>own</u> funds?

It clearly shows that in both periods, Israeli patients pay more out of own funds than German patients and in both periods, differences were found to be significant.

The fact that more and more German counselors work privately causes a conflict between private genetic counselors and hospital institutes (Nippert, 1998). Hospital genetic centers suspect that private practitioners can harm the performance quality

TABLE 3. How Many Patients Pay Out of their Own Pocket?

	1994–96			2001–01		
	Germany	Israel	t-Test	Germany	Israel	t-Test
	Mean (s.d.)	Mean (s.d.)		Mean (s.d.)	Mean (s.d.)	
Own funds	1.27 (0.46)	1.62 (0.59)	−2.619*	1.25 (0.44)	1.82 (0.90)	−3.162**

(1–5 interval scale 1 = 0% 2 = 25% 3 = 26–50% 4 = 51–75% 5 = 76–100%*).*
**Sig* < 0.05 ***Sig* < 0.01

and ethical aspects of counseling services, especially since private physicians can increase their personal profits by offering diagnostic tests. Furthermore, a private practitioner offering the same service as a public clinic nearby may threaten the economic resources of university or hospital-based genetic centers, which are used, as in Israel, to finance research and implement new services (Nippert, Horst and Schmidke, 1997). However, the economic incentive of genetic hospital units in Germany is not very high. This is due to the fact that genetic hospital units in Germany get a fixed payment from the health funds for treatment and thus, their interest in performing extra tests, which may fall back on their limited budget, is reduced.[15] Another option German hospitals have for financing tests is through claiming them to be educational and thus, having the education ministry cover the costs. In this way or the other, German hospitals do not benefit financially from genetic testing; except in rare cases, when testing fees are paid out of pocket or when they are covered by private insurance companies. Hence, those who have the greatest motivation to earn money from genetic testing are private practitioners (mostly gynecologists) and private labs or clinics, which are shared by physicians and laboratory workers who are becoming the main providers of genetic services.

On the other hand, in Israel most counseling and tests are done through genetic hospital units, which benefit from each and every test, usually paid for by the patients themselves (sometimes with the support of the supplementary health insurance) or by the health insurance in cases of a proven high risk. Yet, contrary to common economic thought, free-market principles work rather differently in the two countries. Whereas profit motivations exist in both of them, it seems that commercialization works better through public organization, as is the case in Israel, where genetic hospital units are the main suppliers of genetic prenatal genetic tests but patients pay out of their own pockets. In Germany, three reasons may impede the rapid commercialization of genetic testing, despite the possible profit for private counselors. First is the ethical opposition, which makes it hard for German counselors to offer controversial testing. Second is the fact that private

[15] Information gathered in interviews.

physicians in Germany lack the organizational capacity of big hospital units, their advertising abilities and the trust they receive from patients as public institutions. Third, the health insurance in Germany is more comprehensive than the Israeli one and paying a physician out of pocket, which is nowadays common in Israel, is still rare in Germany.[16] Therefore, German private practitioners find it hard to offer new tests that are not covered by the national health insurance, which unlike in Israel is trusted by the German public to cover all medical necessities.[17] On the other hand, in Israel there exists hardly any ethical opposition to reproductive genetics and Israeli patients are accustomed to not counting entirely on insurance support but rather to paying for "good medicine" out of their own pockets. Therefore, when they are offered tests that are not covered by their health insurance, they do not critically question the offer.

Unfortunately, due to the limitations of the original questionnaire used in this study, and to the fact that almost all of the Israeli counselors work in the public sector, this study does not allow the examination of the effect of working in the private versus the public sector on counselor's practices in both nations. Further studies are needed in order to better understand the influence of work setting and financial motivations on counselors' moral practices. However, from a German-Israeli comparative point of view, it is clear that private financial motivations, which are much stronger in Germany, do not cause the commercialization of this field of medical practice in the Federal Republic, in contrast to institutional profit motivations in Israel, which do lead to stronger commercialization, as I have just explained.

SCIENTIFIC MENTALITÉ, AND ATTITUDES TOWARDS SCIENTIFIC RISKS IN ISRAEL AND GERMANY

It is impossible to understand the cultural logic behind the uses made of medical genetics in both countries, without taking into consideration both countries' scientific mentalité. Being political, moral and aesthetic, the discourse of scientific and medical-genetic risks is profoundly different in Israel and Germany. Whereas Germany is characterized by proliferation of risk discourse and a suspicious attitude towards science and progress, such discourse is almost completely lacking in Israel, where the public is generally trustful of science and "progress" (Golan, 2004).[18] Attitudes towards prenatal genetics therefore cannot be detached from this context.

[16] This has changed slightly since the completion of the study in 2003.

[17] Information collected in the German interviews.

[18] For example, the 2002 survey by the Israeli science, culture and sports ministry concerning public attitudes towards science in Israel, revealed that 73% of the Israeli public was enthusiastic about science, 5.7% was indifferent and only 15% was critical or worried about the outcomes of scientific progress. Moreover, the survey showed that Israelis believe their country has made some impressive scientific achievements in the field of medicine. More than 80% of the surveyed individuals believed that scientists contribute to society and that investment in scientific and technological projects will contribute to Israel's welfare, standard of living and public health. (Public Survey Concerning Science in Israel, 2002).

As for Germany, the Eurobarometer of 2005 http://europa.eu.int/comm/public_opinion/archives/ebs_225_report_en.pdf) shows that in a Europe-wide comparison, most agreement with the statement that "Scientists, due to their knowledge, have a power that makes them dangerous" was found in Malta and Germany. Different writers have pointed to the extraordinary fear of science in the German society and tried to explain its cultural and historical sources. Linke, for example, relates the German preoccupation with scientific and environmental disasters to the German history of the 19th and 20th centuries. She claims that after WWII, Germany had to deal with guilt management, a process in which Nazi values were inverted but their themes, represented in the German collective imagination, were not thoroughly transformed. Linke exemplifies this claim by the fact that the German imagination is still racked with "holocaust" fantasies and fears of mass destruction. The difference is just that now, the German obsession with destruction is directed at nuclear or genetic holocausts, a mass destruction of which this time, the German people would become the victims. Furthermore, Linke argues that the German resentment of science is deeply embedded in the German romantic movement of the 19th century and also in the Nazi universe of the 30s and 40s.

Another commentator on this issue is Maurie J. Cohen, (1999), who claims that there is much in risk society theory, developed by the German social thinker Ulrich Beck, which has a distinctly German spin. Although this theory won international recognition, Cohen suggests that its central contention that public anxieties about high-consequence low-probability events undermine the legitimacy of science, best describes German and not modern mentalité. Her explanation of the German specific scientific mentalité, has to do with German religious doctrine, moral prescripts, historical precedent or simple inertia. Therefore, she reads the famous Ulrich Beck's characterization of the contemporary era as "risk society" as a German-inspired approach, whose relevance for other Western societies demands greater scrutiny. Adopting her advice, I argue that the Israeli society does not fit the theoretical model of a risk society and that the relationship between science and society in Germany and Israel are dramatically different. Beck's thesis is that contemporary Western society is no longer occupied with the production and distribution of scarce goods but with the prevention or minimalization of "risks", which are the outcome of modernization and industrialization. However, Beck tends to downplay the fact that what concerns one contemporary modern society may not concern another, as he stresses that risks are objectively heightened in the late modern era. In opposition to Beck, I do not claim risks are objective,[19] and hence what I stress is the differences between contemporary societies and not between pre-modern and late modern ones.[20] Concretely, I argue that Germans are occupied with probable distant risks, which are hard to measure because they are not localized and supposedly have a long-term effect, whereas Israeli worries are far more tangible (actually fearing

[19] Beck himself sometimes seems to claim that risks are objective and other times to be more critical of this claim. Yet, to my mind his overall thesis does treat scientific risks as more real than imaginary.

[20] For a discussion of whether "Risk society" or Re-modernization is happening at all, see Latour, 2003.

the next terror attack, or war). In that sense, Israel may have not reached the stage of a late-modern risk society, which is centered on critical self-reflection upon the dangers of modernity. On the other hand, Germany, which experienced Nazism, which Bauman (1998) understands to have been a typical modern social engineering regime in which science was uncontestedly expected to create a better, more orderly and healthier society, is now no longer innocent concerning the atrocities of the modern rational-scientific spirit. This can be best demonstrated by the traditional lack of criticism concerning nuclear power in Israel. Whereas the fear of the all-destructive atom bomb was behind the formation of environmental oppositional groups in much of the Western world and certainly in Germany, Israel seems to be missing from this line of protest, which later on formed a coalition with environmental concerns about gene technology in Germany and elsewhere. As a matter of fact, when "Greenpeace" International opened their office in Israel, they decided to leave aside the nuclear power protest, as they estimated it would not afford them any sympathy among the Israelis, who basically feel protected and not endangered by their own nuclear abilities both because of the Zionist-Palestinian conflict and because of their trust in science. Speculating about the reasons for these differences between Germany and Israel is beyond the scope of this work. Yet the immediate line of thought that comes to mind has to do with the fact that in the last sixty years Israel has been faced with very concrete risks, whereas Germany has been a peaceful state. Returning to genetics, all forms of gene technology (in medicine, agriculture or ecology) are viewed skeptically by the German public. In fact, according to a European survey done in 1991, there is no other country in the world where gene technology has encountered resistance as strongly as in the former West Germany (Abbott, 1996; Dixon, 1993). Dixon's survey, which assessed how genetic engineering scientists in Europe view the influence of public attention on their field, found out that in almost every respect, by far the most negative responses came from scientists in West Germany, who felt deeply criticized by the public. German scientists and journalists who were asked to explain this phenomenon pointed to three causes for public opposition: one-sided media coverage, sensitivity to the WWII history of eugenics and powerful activist opposition. Furthermore, the "Eurobarometer" 1991 and 1993 surveys of the German general public, showed that the German scientists' perceptions of the public attitudes were based on reality, as the application of genetic technology in Germany has been continually countered by presentations of ethical, legal, socio-economic and environmental risks (Rabino, 1994). However, this tendency seems to be changing, as recent studies suggest that the German public favors allowing new genetic reproductive technologies, despite the restrictive German laws concerning this issue (Krones, 2005).

Writing about the regulation of biotechnology in Israel and its non-controversial status as compared to other parts of the Western world, Prainsack and Firestine (2006) argue that Israelis generally adopt a positive attitude to scientific practices and technologies that are controversial elsewhere (such as stem cell research, genetic diagnosis, and cloning). This attitude is explained in terms of cultural, political and religious narratives, which construct biotechnology as crucial for the continuity of Jewish existence within the Middle East; indeed, the very survival of Israel in such a hostile environment is seen to be dependent on its modernity or in other words, on its scientific and technological superiority (Prainsack and Firestine, 2006). To get back to the specific interests of this study, let me remark that although the German genetic counselors I studied were critical of the fears of science and genetics in their society, those fears and their language and metaphors penetrated their own expert discourse and affected their attitudes towards their expertise. Israeli counselors on the other hand, were uncritical of science, just like the public they belong to.

CONCLUSION

This chapter has been dedicated to a comparative overview of the Israeli and German prenatal genetic fields' history, budgeting, cultural-religious contexts and political logic. In it I have discussed different actors involved in the field of reproductive genetics in Israel and Germany, such as the field's experts, the field's supporters and opponents, the field's earners, such as private and public clinics that offer prenatal genetic tests and above all, the Israeli and German states, with their unique medical establishments, history and regulatory laws. My leading question was: what is it about the fields' history, contemporary organization and cultural environment that can shed light on the dramatic differences in the adoption of reproductive genetic practices between the societies of Israel and Germany.

Regarding history, I showed that the start of applied human genetics in Israel after the establishment of the state in 1948 was not controversial, as opposed to the situation in Germany after WWII. I further claimed that the lessons learned from Nazi history have a strongly restrictive effect on the development of reproductive genetics in Germany up to present, whereas in Israel this history has been forgotten or rather replaced, by the Jewish-Zionist "soft" eugenic tradition, which is in line with today's practices. Regarding the possible effect of different occurrence rates of birth defects in the two nations, I claimed that the differences between the two fields can hardly be ascribed to such dissimilarities. Furthermore, I demonstrated that conventional economic logic cannot explain the differences between the two states and the fact that the commercialization of genetic tests in Israel exceeds by far the comparatively non-commercialization of tests in Germany, since: A. Health insurance coverage is more inclusive in Germany than in Israel, both generally and specifically for reproductive genetic examinations and yet, it does not lead to the expansion of human genetics in Germany. B. Private profit motivations are stronger in Germany but counter to prevalent economic reasoning, this does not push medical genetics forward in Germany. Concerning the role of the two states, I showed that the Israeli state, as well as its medical institutions, clearly supports the prevention of "birth defects", not just on a private level but also on the community level, whereas the German state's, and even more so the German health establishment's, encouragement of prenatal genetic diagnosis and its derivatives is far more limited. Moving to the public's compliance with such tests and their common outcomes,

I demonstrated that no large differences exist between Israeli and German women in their acceptance of repro-genetics and that such differences cannot explain the studied question. I furthermore claimed that the existence of a loud opposition to reproductive genetics in Germany, contrary to its non-existence in Israel, is a very crucial factor behind the actual differences in the uses of reproductive genetics in both sites and that religious traditions play a major role in shaping what is morally thinkable. Additionally, I showed that professional ethics differs in the studied nations, despite the fact that they belong to a global professional network, as in Germany many more restrictions are put on genetic counselors, (for example by their own guidelines) and as professional-ethical discussion is flourishing in Germany as opposed to Israel, where it hardly exists. The overall German critical discourse of science and specifically of genetics, as opposed to the non-critical Israeli discussion of such matters, were the last factors I described as shaping the culture and structure of the fields of reproductive genetics in both sites.

To conclude, this chapter has demonstrated that biopower over life and death is exercised in both countries in a dissimilar fashion. Hence, from the pregnant women's, or the public's perspective, one can question just how voluntary genetic tests really are in Israel, since patients are seduced into taking exams by the generally supportive atmosphere for such further medicalization and commercialization of pregnancy. From the German point of view, German women – and the German public in general – tend to be deprived of information about medico-technical innovations, which they lack the chance to choose or decline and which affect their supposedly autonomous life and death decisions.

The following chapter will portray the major findings of the genetic counselors survey results in Israel and Germany and discuss the effects of the counselors' cultural-national, institutional and personal backgrounds on their attitudes to different issues such as abortions and selective abortions, eugenics, disability, patients' autonomy, and the rational-scientific planning of life.

CHAPTER 4

GENETIC COUNSELORS' MORAL PRACTICES

WHY STUDY GENETIC COUNSELORS?

Enabling the discovery of genetic characteristics of the embryo/fetus, the medical field of reproductive genetics is currently the largest consumer of genetic technology. However, because it does not (yet?) allow the "curing" of embryos who will (probably) suffer from a genetic condition in the future, anomalous findings become matters that need to be decided upon, (Beck and Beck-Gernsheim, 1995). Indeed, the common decision in such cases takes the form of what has become known as "selective" abortion.[1] This chapter focuses specifically on the moral practices of genetic counselors in both countries and on the social factors behind them.

Why focus on genetic counselors? Since state-coerced eugenics is no longer practiced in "advanced liberal" societies, when a "positive" genetic diagnosis is made (or suspected), it is the future parents who make the delicate decisions concerning the interruption or continuation of the pregnancy. However, they do not do so alone. Rather, they consult genetic counselors, who are the bearers of knowledge and expertise in this field (Ettorre, 1999) and thereby among the major producers of genetic ideology. Through their use of metaphors, stories and writings, they articulate, construct and reproduce their position of authority and fulfill their role as interpreters of knowledge (Ettorre, 1999). Thus, genetic counselors, namely doctors and biologists who practice genetic counseling, are among the most prominent "gate-keepers" in shaping parents' decisions about their future children and standards of entry into the human community (Rapp, 1999). Equally, being a well-organized professional group, genetic counselors also strive to shape public policy regarding prenatal diagnosis (Wertz and Fletcher, 2004). In this regard, genetic counseling is a typical case of expert knowledge and advice, which are pivotal to what Foucault termed normalization, governmentality (Foucault, 1991) and biopower, (Foucault, 1978). As such, they provide the guidelines and council by which populations are screened, compared against norms and rendered productive, and by which late-modern subjects and their bodies are governed and fabricated within a network of instruments and techniques of power.

Yet, saying this, I am not implying that counselors' opinions are necessarily in line with those of the public or with the positions of specific sub-groups within the general public. For example, in Germany, Irmgard Nippert and Gerhard Wolff (1999)

[1] Such abortions are also often termed "medical", "therapeutic" or "eugenic". These terms relate to wanted pregnancies, which are interrupted due to embryopathic indications.

have shown that German patients are much more inclined than German geneticists to support the idea that people have a duty to prevent birth defects or that people have a responsibility not to pass on serious genetic diseases or disabilities to their children.

Furthermore, Erikson (2003) has shown that when faced with a "positive" diagnosis, most German women opt for selective abortion. In that sense, the power that professionals hold restricts the possible uses of medical genetics in Germany, as it forms a barrier between what is desired by patients and what is offered to them. On the other hand, different studies have demonstrated that health professionals in Israel are significantly more supportive of comprehensive prenatal testing than women, the potential consumers of the tests, especially those coming from Orthodox Jewish and Arab communities (Mishori-Dery et al., forthcoming; Sher et al., 2003). Hence, it is also important to understand counselor's moral practices in order to understand potential conflicts between them and their clients.

A BRIEF HISTORY OF GENETIC COUNSELING AND ITS NON-DIRECTIVE ETHOS

As this chapter deals with counselors' moral practices, a brief history of the profession and its ethical ethos is in order. Genetic counseling began in the late 1940s, soon after World War Two; previously, in the 1930s, it had been called "genetic hygiene" or "genetic advice". Eugenic practices were widely and infamously used in Nazi Germany, though they were also widespread in the US, Sweden and other countries (Paul, 1992), as well as in the "*Yishuv*", the Jewish community in Palestine (Falk, 2002; Hirsch, forthcoming; Stoler-Liss, 1998). In 1947, the term "genetic counseling" replaced eugenically tainted terms (Reed, 1974) and in 1974 a committee of the American Society of Human Genetics suggested "nondirectiveness" as the norm for genetic counseling (Fraser, 1974),[2] thus diverging from all other medical specialties (Wertz, 1997).

The principle of nondirectiveness is based on the ethical principle of respect for autonomy, which refers to the individual's right to liberty and privacy in making decisions. Ethically speaking, the responsibility for genetic decisions is transferred from the professionals (genetic counselors) to the patients, whose autonomy is now glorified. Wolff and Jung (1995) offer three explanations for the existence of the nondirective principle in genetic counseling. Firstly, it was initially biologists, who were unfamiliar with the paternalistic concept of doctor-patient relationships, who practiced genetic counseling and not physicians. As a professional group, they were strongly influenced by concepts connected to psychological counseling, especially

[2] The classic meaning of nondirectiveness, as described by Fraser in 1974, includes the following elements: not making decisions for your patients, supporting their decisions (whatever they are), helping couples and individuals understand their options and the present state of medical knowledge so they can make informed decisions, helping individuals/couples adjust to and cope with their genetic problems, the removal or lessening of patient guilt or anxiety and helping individuals/couples to achieve their parenting goals.

the client-centered therapy of Carl Rogers, which promotes helping the client to reach her/his own solutions. Secondly, Wolff and Jung refer to the need to avoid conflict. The fact that genetic counselors invoke the principle of nondirectiveness can be viewed as an attempt to avoid confronting the conflicting objectives and ethical principles in the field. As a result, such conflicts are not dealt with in either general or individual cases. Thirdly, the affirmation of nondirectiveness serves as a defense against attacks on the allegedly harmful nature of applied human genetics, as it prevents critical discussion of both the psychosocial and the societal implications of genetic knowledge. Furthermore, it must be remembered that in the early days of genetic counseling, it had no treatments or even preventive measures to offer, apart from avoiding pregnancy (Wertz and Fletcher, 2004).

While the non-directive practice of genetic counseling is usually referred to as value neutral, Caplan (1993) convincingly argues that "the shift toward an ethic that elevates client or patient autonomy above all other values is highly value laden and prescriptive" (Caplan, 1993, pp. 159–160). In similar fashion, Clarke (1991) argues that even if practiced in a non-directive manner, PND in and of itself establishes a directive context and serves to transfer moral responsibility to patients, while helping counselors wash their hands of it. Clark writes: "I contend that an offer of PND implies a recommendation to accept that offer, which in turn entails a tacit recommendation to terminate a pregnancy if it is found to show any abnormality" (Clarke, 1991 p. 1000).

The genetic counseling ideal of nondirectiveness, typical of the widespread American professional culture of value neutrality that works to conceal political conflicts and present them as apolitical,[3] is prevalent in both Germany and Israel and can be viewed as characteristic of the global influence on professionals, at least in the West. However, while professional global ethics, such as that of nondirectiveness, supposedly push societies throughout the world towards similar moral practices, local morality (Kleinman, 1995) may well be pushing in the opposite direction. Studies by Wertz and Fletcher (2004) and Van Zuuren (1997) show that outside the English-speaking world, most counseling is actually directive, as the counselors provide their clients with intentionally slanted information in order to influence their decisions. (Clarke, 1991). The explanation for this is familiar: the spheres of culture and science cannot easily be separated and medical knowledge is continuously permeated by culture (Kleinman, 1995; Lippman, 1991; Lock, 1993; Martin, 1991; Payer, 1988; Wright and Treacher, 1982).

This study focuses on the testimonies of counselors concerning their own moral practices. It shows that despite the non-directive norm of counseling that is prevalent in both societies, barely a single respondent systematically answered that they would try to be as non-directive as possible in counseling for 26 different abnormal conditions during pregnancy. In fact, in line with Fine's arguments (1993), my

[3] For other discussions of professionals' supposed value neutrality see, for example, Shenhav's book on the American managerial revolution. Yehuda Shenhav, *Manufacturing Rationality: The Engineering Foundations of the Managerial Revolution* (Oxford, 1999).

findings demonstrate that the ethos of neutrality actually exists alongside a spectrum of beliefs, attitudes and conduct. While previous studies have shown that genetic counseling varies across advanced societies (Van Zuuren, 1997; Wertz and Fletcher, 2004), none of them have looked deeper into this variation and examined the influence of cultural, professional and personal factors on counselors' practices in dissimilar nations. Hence, in the following I will first present the nature of the differences found between the groups studied. Second, I will ask what factors affect the counselors' moral practices and whether it is cultural, professional or personal characteristics which mostly affect their beliefs and conduct. These findings will be discussed from a number of theoretical perspectives: the global-local perspective, a feminist perspective and the theory of technological determinism. But first, let me briefly mention whom is allowed to council in both states.

WHOM IS ALLOWED TO COUNCIL?

In Israel, according to The Israeli Law of Genetic Information (2000), four professional groups are allowed to practice genetic counseling: 1. M.D. medical geneticists; 2. Clinical geneticists: holding a Ph.D. in the life sciences with 2 years training in genetic counseling, done in a recognized genetic institute or laboratory; 3. Genetic counselors: holding an MA in human genetics, with practical training in genetic counseling, done in a recognized genetic institute or laboratory;4. When the genetic problem is related to other medical conditions, MD's with diverse medical specializations are allowed to counsel for matters related to their special expertise. Since 1986, medical genetics is a specialty recognized by the Israeli Medical Association and the Ministry of Health (Zlotogora and Chemke, 1995).

In Germany, three professional groups are allowed to counsel:

1. M.D.'s with a "special medical education" (Zusatzbezeichnung) in genetics, which was acknowledged in 1978 by the certification boards of the states; 2. M.D.'s with a specialty in human genetics (Facharzt für Humangenetik). This specialization has been offered since 1992. It requires a training period of 5 years in all fields of human genetics. The difference between the two recognitions is important: *Zusatzbezeichnung* (special medical education) was the first step to professional recognition but it took another 14 years for human geneticists to be recognized as a medical specialty, which is an officially recognized area of medical practice (Nippert, Horst and Schmidke, 1997); 3. Another professional group that is allowed to counsel are non-M.D's, holding a Ph.D. in human genetics (Fachhumangenetiker). This specialization was also recognized in the 90s.

SURVEY DATA: METHOD OF ANALYSIS

(for data and sample see Chapter 2)

Firstly, I examined the degree of support among German and Israeli genetic counselors for selective abortion for each of the 26 different genetic conditions included in the survey, such as CF, anencephaly, sex undesired by parents, trisomy

13, cleft lip and palate, Down's syndrome, sickle-cell anemia and the like (for the full list see Appendix 1). Systematic comparisons of the two groups were conducted by means of the T-Test procedure for independent samples.

Secondly, in order to ensure that the differences between the two groups were not masked or enhanced by factors other than nationality and to check for the net effect of independent personal, professional and cultural variables on different moral issues related to reproductive genetics (such as abortion in general, eugenics, attitudes towards the disabled, patient autonomy and trust in science), I used linear regression equations.

Variables

Dependent Variables: at the focus of the research are counselors' practices concerning selective abortion. In order to measure them, counselors were asked to state their level of support for the termination of pregnancy for each of 26 different genetic conditions.

Six dependent variables were used in the regression analysis. The first is an index based on answers concerning counseling for the above-mentioned 26 genetic conditions. The other dependent variables are indexes built on the average of answers to a cluster of questions concerning different issues that are commonly understood to be relevant to genetic counselors' moral practices (see Appendix 4 for indexes' descriptive statistics and reliability tests). Among them are social-moral dilemmas provoked by the practice of PND, as well as prevalent medical-ethical norms.

The variables used were ordinal (mostly checking for degree of agreement or disagreement with regard to different practices and opinions). Three groups of independent variables were used: cultural (consisting in nationality and religiosity), professional (consisting in place of training, medical subspecialty and years on the job) and personal (consisting in sex and age) For full operational definitions of the variables see Appendix 3. Another independent variable was time of the survey.

Findings

Table 1 shows that in 1993–95, there was an almost equal number of men and women among genetic counselors in both Germany and Israel. In 2000–01 on the other hand, due to the inclusion in the second sample of non-MD genetic counselors, women made up the majority of surveyed counselors in Israel (see Table 1, remark 1). Israeli counselors were slightly older than their German counterparts in both samples, though the difference was significant only in the first sample, a fact which may be due to the slow start of medical genetics in Germany after the war (Deichmann, 1996). Furthermore, most Israeli counselors were Jewish (100% and 93.3%), whereas most of the German counselors were Christian (71.2% and 76.8%, of whom roughly a quarter were Catholic). In addition, German counselors were significantly more religious than their Israeli counterparts in both periods. No systematic differences were found between Israeli and German counselors in their years of professional experience. In both periods, 22.8% and 33.3% of German

TABLE 1. Descriptive Data about the Counselors in Both Countries

	1993–5			2000–01		
	Germany $N = 185$ Mean (s.d.)	Israel $N = 22$ Mean (s.d.)	t-Test	Germany $N = 57$ Mean (s.d.)	Israel $N = 31$ Mean (s.d.)	t-Test
% Males (1)	48.6%	50%		50%	25.8%	
Age (in years)	43.63 (8.55)	49.18 (8.85)	−2.871*	48.47 (9.46)	45.33 (9.98)	1.445
% Jews (2)	0.5%	100%		1.8%	93.3%	
% Protestant	42.9%	–		51.8%	–	
% Catholic	28.3%	–		25%	–	
% Muslim	0.5%	–		–	–	
% None	27.7%	–		19.6%	6.7%	
Importance of religion (3)	2.37 (0.99)	1.73 (0.70)	−3.844**	2.56 (1.04)	1.80 (0.89)	−3.419**
Years in medical genetics	11.35 (7.52)	12.09 (9.49)	−0.426	14.2 (8.80)	10.69 (9.92)	1.616
% who work in private (as opposed to public) settings (4)	22.8%	4.5%		33.3%	3.4%	
Trained only in home Country	89.5%	31.8%		78.9%	48.4%	
Trained also in the US	4.3%	54.5%		10.2%	45.2%	
Pediatricians	28.6%	40.9%		22.4%	38.7%	
Gynecologists and Obstetricians	14.6%	13.6%		6.9%	3.2%	
Non-MD g. counselors (5)	0%	0%		1.7%	48.4%	

* *Sig* < *0.05* ** *Sig* < 0.01.

(1) In 1993–95 only MD genetic counselors were included. Women constitute about 1/3 of this population in Israel but they are over-represented in the sample. In 2000–01 non-MD counselors were also included. Women make up more than 95% of this population (in Israel). This partly explains the over-representation of women in the second sample in Israel. Non-MD counselors are hardly present in the German sample, since they were only allowed to counsel in Germany shortly before the survey was conducted. (2) In Israel, stating an affiliation to the Jewish religion is used interchangeably with affiliation to the Jewish people. Therefore the "None" category is problematic. Most secular Jews would answer "Jewish", despite being non-religious. (3) The question asked was: How important is religion in your life? (1–5 interval scale: 1-Not at all important, 2-Slightly important, 3-Moderately important, 4-Very important, 5-Extremely important) (4) Genetic counseling in Germany has historically been located in public settings, which is quite unique in the overall German medical system, as German physicians either work entirely in private practice or entirely in a public setting. (see Chapter 3). Currently, genetic counseling in Germany is moving outside public settings to private clinics. No equivalent statutory dichotomy between the two sectors exists in Israel. Thus, the majority of Israeli counselors conduct most of their work in public settings but may also run a private clinic. (5) In Germany, non-MDs (Fachhumangenetiker, who are non-MDs but with a PhD in human genetics) were only recently allowed to counsel patients. In Israel, non-MD counselors comprise a large part of the professional group, which resembles the American system.

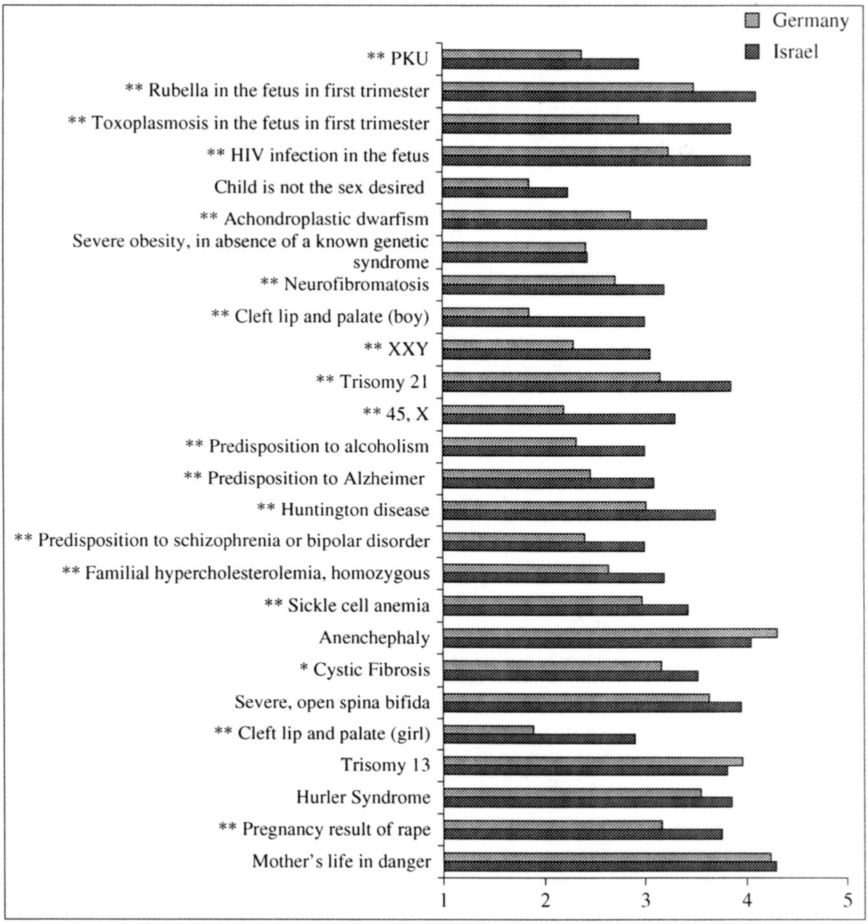

* Sig < 0.05 ** Sig < 0.01

Figure 1. How would you counsel? for and against abortion 1993–1995

counselors worked only in the private sector, as opposed to the fewer Israeli counselors, who mostly conducted their work in the private sector (4.5% and 3.4%, see remark 4). The majority of German counselors in both periods were trained only in Germany (89.5% and 78.9%), while many of the Israeli counselors had also trained abroad (68.2% and 51.6%), mostly in the US (54.5% and 45.2%). Pediatricians made up 40.9% and 36.7% of Israeli genetic counselors but only 28.6% and 22.4% of the German counselor population. Furthermore, gynecologists and obstetricians constituted 13.6% and 3.2% of the Israeli counselors and 14.6% and 6.9% of the German group. Non-MD counselors did not exist in the German sample, whereas in the second sample in Israel, they made up 48.4% of the group (see remark 5).

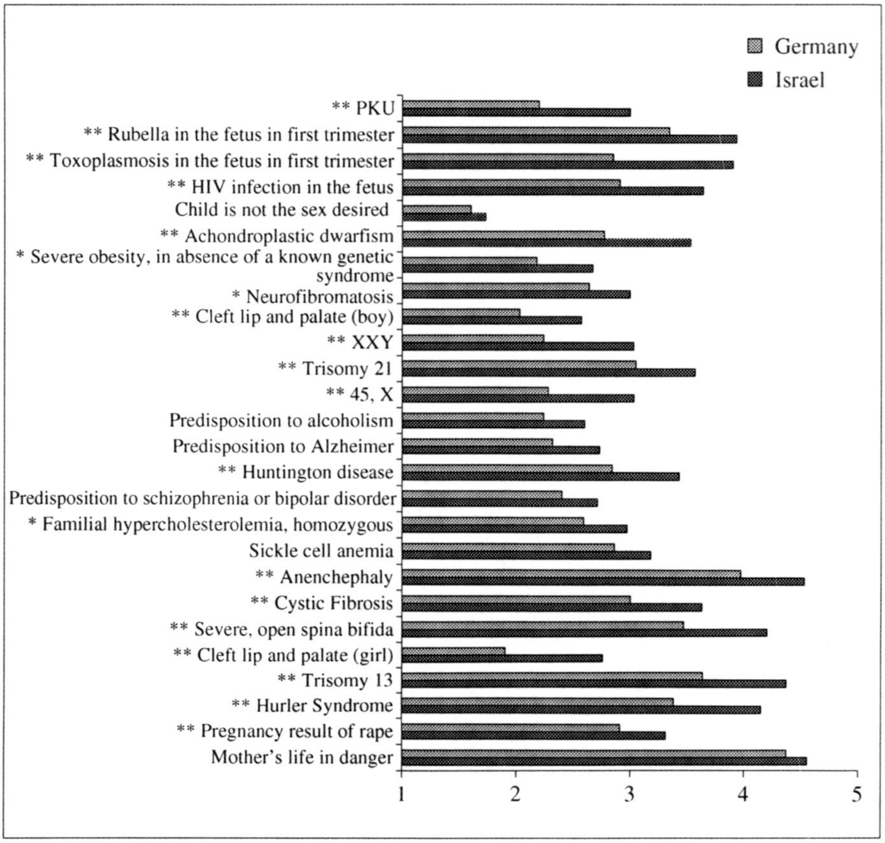

*Sig < 0.05 ** Sig < 0.01

Figure 2. How would you counsel? for and against abortion 2000–2001

Concerning differences in professional practices and norms between the two studied groups, the following figures (Figures 1 and 2) describe the answers given by genetic counselors in both countries and in the two time periods, to the question: **How would you counsel about termination of pregnancy for the following conditions?** (*Options for answers: 1. Urge to carry to term 2. Emphasize positive aspects 3. Be as unbiased as possible 4. Emphasize negative aspects 5. Urge termination*). (For complete tables see Appendix 2.)

According to Figure 1, in 1993–95, when asked how they would counsel following prenatal diagnosis, Israelis were significantly more supportive of abortion than Germans in 19 out of 26 different fetal conditions. In the other seven cases (mother's life in danger, Hurler Syndrome, trisomy 13, Severe open spina bifida, anenchephaly, child not the sex desired by parents and severe obesity), no significant differences were found between the groups.

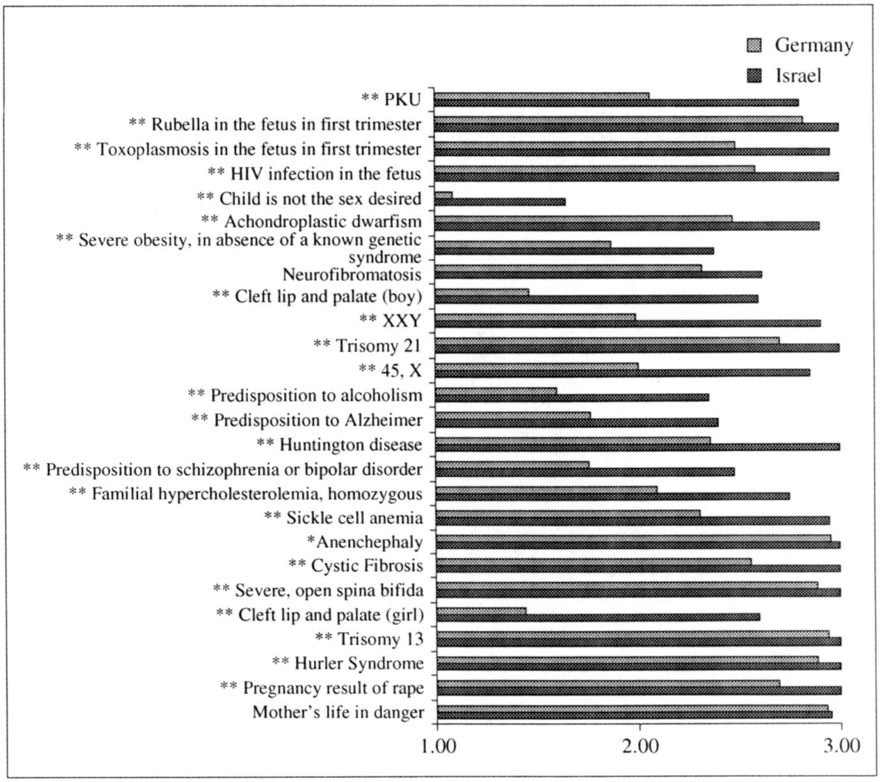

Figure 3. How would you act personally (1st trimester)? 1993–1995

In 2000–01 (Figure 2), Israelis were significantly more supportive of abortion in 20 out of the 26 conditions. In the other six cases (mother's life in danger, sickle-cell anemia, predisposition to schizophrenia or bipolar disorder, predisposition to Alzheimer's disease, and child is not the sex desired by the parents), Israelis were more supportive of abortion but the differences were not significant (see Appendix 2).

Assuming that personal bias influences professional counseling (Nippert and Wolff, 2004), the counselors were also asked: **"How would you personally respond if you yourself in your own life faced the possibility of having a child with the disorders listed below?"** *(Options for answers 1. I would have an abortion 2. I would not have an abortion, but it should be legal for others 3. I would not have an abortion and it should be illegal for others).* All questions refer to the first 3 months of pregnancy, in order to avoid the "late abortion" conflict. (The order of the numbers of the answers was reversed in the figures).

Figures 3 and 4 describe the counselors' answers in 1993–95 and 2000–01 respectively. In 1993–95, Israelis were significantly more in favor of abortion in 24 out

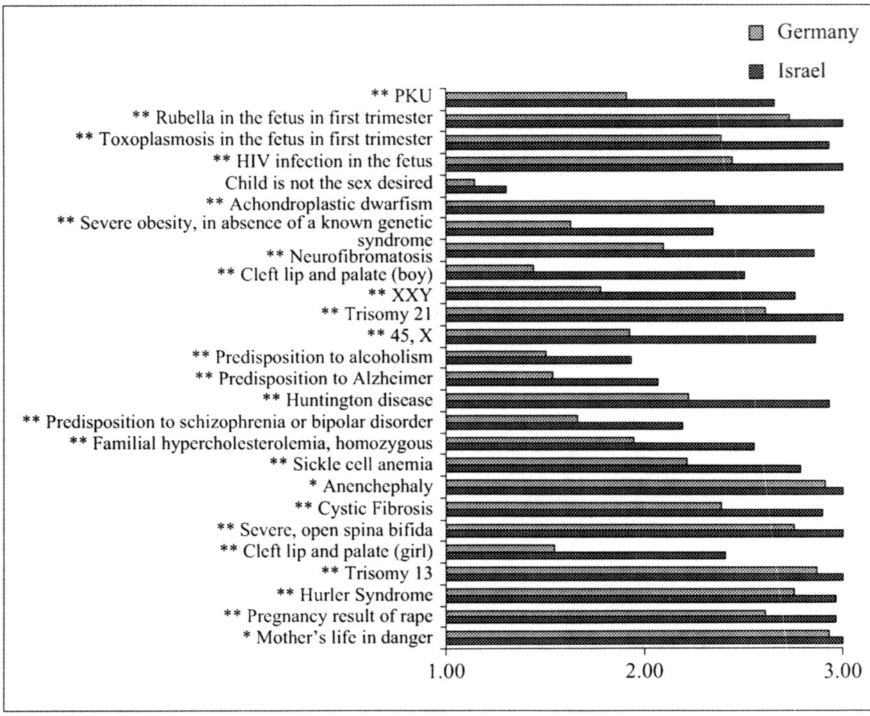

Figure 4. How would you act personally (1st trimester)? 2000–2001

of 26 cases. In the other two cases (mother's life in danger and neurofibromatosis) no significant differences were found. In 2000–01, Israelis were significantly more supportive of aborting their own fetus in all but one case (child not the sex desired by the parents).

SELECTIVE ABORTION AND OTHER SOCIAL-MORAL DILEMMAS

So far, we have seen that systematic differences were found between Israeli and German genetic counselors in both their professional practices and expected personal behavior in the vast majority of the 26 genetic conditions referred to in the questionnaire. Attitudes towards selective abortion are commonly related to social-moral dilemmas and issues provoked by the practice of PND. These include: the moral status of embryos and accompanying attitudes towards abortion in general (German National Ethics Council, 2003); a wish to detach oneself from eugenics and its immoral connotations (Duster, 1990; Koch, 2004; Paul, 1992); a fear of devaluing the life of the disabled (Asch and Geller, 1996; Blumberg, 1994; Parens and Asch, 2000; Ward, 2002); a demand to respect patients' autonomy and their right to services (Wertz and Fletcher, 2004) and trust and faith in the rational-

TABLE 2. Correlation matrix between attitudes towards selective abortion and attitudes towards the social-moral dilemmas provoked by the practice of PND

	Pro Selective Abortions	Pro Abortion (in general)	Pro Eugenic Practices	Pro Disabled are Good for Society	Pro Parental Autonomy	Pro Rational-Scientific Planning of Life
Pro Selective Abortions	–	0.318**	0.192**	−0.455**	0.275**	0.411**

* < *0.05*
** *0.01*

medical-scientific planning of life, which is not often discussed in writings on this issue but which I found relevant to German and Israeli attitudes, due to the different scientific mentalities of both societies (see Chapter 3).

In order to measure counselors' attitudes towards the moral quandaries prevalent in their field and relevant to their practice (i.e., attitudes towards non-selective abortion, eugenic practices, the disabled in society, parental autonomy and the rational-scientific planning of life), indexes based on the averages of answers to a cluster of pertinent questions were constructed (see Appendix 4).

Fairly strong correlations were found in a correlation matrix between attitudes towards selective abortion and attitudes towards the aforementioned social-moral dilemmas (see Table 2). Specifically, it was found that the higher the support for selective abortion, the lower the support for the idea that the disabled are good for society and the higher the support for abortion (in general), eugenic practices, parents' autonomy and the scientific-rational planning of life. Moreover, significant differences were found between the two national groups with regard to the aforementioned social moral dilemmas, except the one which concerns eugenic ideas (see Appendix 5).

Hence, in the next stage of the study, a regression analysis was designed to check for the influence and relative importance of cultural, professional and personal factors in shaping counselors' practices. In addition, the effect of time was examined. (see Table 3).

WHAT DETERMINES COUNSELORS' MORAL PRACTICES?

Discussion

The findings (see Figures 1–4) suggest that Israeli professionals counsel and would also personally choose, for abortion of embryos (and fetuses) with imperfections that in Germany are not looked upon as justifying the termination of a pregnancy. Thus, at the prenatal stage, non-tolerance towards the genetically deviant is the norm among Israeli counselors, while more tolerant or optimistic counseling characterizes German counselors.

TABLE 3. Variables Predicting Prenatal Diagnosis Practices: Regression Results of Both Samples (1993–95 and 2000–01): (For description of independent variables see Appendix 3)

PND Practices / Independent Variables	Pro Selective Abortions B (S.E.)	Pro Abortion B (S.E.)	Pro Eugenic Practices B (S.E.)	Pro Disabled are Good for Society B (S.E.)	Pro Parental Autonomy B (S.E.)	Pro Rational Scientific Planning of Life B (S.E.)
Germany[a]	−0.351**	−0.402*	−0.066	1.163 **	−0.579**	−0.681**
	(0.045)	(0.172)	(0.127)	(0.128)	(0.090)	(0.089)
Female[b]	−0.029	0.236 *	0.111	−0.023	0.108	−0.164**
	(0.030)	(0.113)	(0.085)	(0.086)	(0.056)	(0.060)
Age	0.003	0.017	0.006	−0.006	0.007	0.005
	(0.003)	(0.011)	(0.008)	(0.008)	(0.005)	(0.005)
Religiosity	−0.026	−0.194**	0.001	0.037	−0.043	0.036
	(0.015)	(0.055)	(0.041)	(0.041)	(0.027)	(0.029)
Trained only[c] in home country	0.045 (0.040)	−0.187 (0.152)	0.120 (0.117)	−0.120 (0.114)	−0.341** (0.079)	−0.015 (0.081)
Years in Medical Genetics	−0.00005 (0.003)	0.002 (0.012)	0.003 (0.009)	−0.003 (0.009)	0.001 (0.006)	0.0007 (0.006)
Pediatrician[d]	0.054	−0.138	0.247**	−0.046	−0.026	0.102
	(0.034)	(0.127)	(0.096)	(0.097)	(0.064)	(0.067)
Obstetrician[e]	0.018	0.044	0.184	−0.163	−0.054	−0.038
	(0.046)	(0.171)	(0.126)	(0.130)	(0.086)	(0.090)
2000–01[f]	−0.008	0.003	−0.378**	0.309 **	0.110	−0.206 **
	(0.035)	(0.131)	(0.097)	(0.099)	(0.066)	(0.069)
Constant	3.242 **	3.239 **	1.635 **	2.906 **	3.256 **	3.072 **
	(0.116)	(0.439)	(0.328)	(0.334)	(0.216)	(0.228)

* *< 0.05*
** *0.01*
[a] *Reference category – Israeli counselors.*
[b] *Reference category – Women.*
[c] *Reference category – Trained abroad.*
[d] *Reference category – All other MD's.*
[e] *Reference category – All other MD's.*
[f] *Reference category – 1993–95.*

Differences between the two groups tended to be significant, especially regarding the question of counselor's personal behavior. In fact, differences between German and Israeli counselors' personal choices were non-significant in only two cases, that of danger to the mother's life (in both surveys) and that of neurofibromatosis (in the first survey). Thus, while Israeli counselors would choose abortion, were they to personally deal with the dilemma of selective abortion, far more than their German counterparts, abortion was very strongly supported by both groups in the case of danger to the mother and hence, differences in that regard are exceptionally small and insignificant. However, when the decision centers on the prospected health of the child to be, differences between the two groups are very dramatic.

Regarding counseling, again, in the majority of cases presented to counselors in both surveys, support of abortion was significantly stronger among Israeli counselors. However, in the second survey (Figure 2) differences were not significant in some controversial cases: those involving sex selection and predisposition to late onset genetic conditions, such as alcoholism, Alzheimer's disease and schizophrenia. Likewise, differences concerning sickle-cell anemia and danger to the mother's life were non-significant.

In the first survey (Figure 1) non-significant differences between the two groups of counselors were characteristic of either cases in which support for abortion was rather weak among both groups (such as child's sex and severe obesity in the absence of a known genetic syndrome) or in cases in which support for abortion was rather strong among both groups (such as anenchephaly and trisomy 13, in which German counselors' support of abortion was even higher than that of Israeli counselors, though not significantly, as well as severe open spina bifida, Hurler syndrome and mother's life in danger). Thus, it seems that extreme cases, in either the perceived severity of the genetic condition or in their moral acceptability, seem to bring opinions closer together.

THE EFFECT OF NATIONALITY

Modern genetics is a global enterprise (Wertz and Fletcher, 2004) and genetic counselors in both Germany and Israel are part of a scientific community which, due to its practices of interaction (academic literature, international conferences and the like), is global by definition. Belonging to such a network of international scientists, the counselors in both societies share up-to-date scientific knowledge about embryonic development and pathology. Outdated social theories, such as those of Parsons and other functionalists, claimed that professionals have a value system and behavioral patterns that differ from those of other occupational groups (Goode, 1957). Diverse studies subsequently proved this to be only partially true, as professionals were shown to be affected both by their profession and by their cultural setting (Goldberg, 1976). In today's global world, the local/global interplay is a key scenario in social theory (Appadurai and Breckenridge, 1988; Morley, 1991; Robertson, 1993, 1995; Tomlinson, 1999), indicating mutual influences between the two. However, it is still often thought that professionalism is an important factor pushing towards a global orientation that harmonizes practices (Featherstone, 1995; Goldberg, 1976). Hence, our first question concerns whether German and Israeli genetic counselors' moral practices are first and foremost global or local and whether their understandings of different genetic conditions and professional recommendations emanate primarily from their socialization into a particular scientific discipline or rather, from their socio-cultural background. The contemporary discussion concerning globalization processes introduces the concept of glocality (Ram, 1999; Robertson, 1993, 1995). Combining the terms "global" and "local", glocality describes a post-modern state of affairs in which global and local tendencies are interwoven in a dialectical and non-binary process (Gooldin, 2002). In this context, we must ask how genetic counselors, who are positioned in both a global and local world operate, and how professional norms are adopted into and adapted by local cultural-meaning systems.

Both Goode (1966) and Moore (1970) have claimed that professionals' local (as opposed to cosmopolitan) orientations may be described as related to "service orientations", or to their "down-to-earth" contact with clients, as opposed to abstract ethical orientations. Writing about medical anthropology, Kleinman (1995) pointed to the distinction between what he terms the "ethical" and the "moral". Whereas the "ethical" is a codified body of abstract knowledge held by experts about "the good" and the ways to realize it, the "moral" is embedded in a local context and refers to the commitments of social participants in their everyday experience. Modern medicine as a body of knowledge is abstract, scientific and diffuses easily across national boundaries. However, it should be remembered that this knowledge is always applied clinically by professionals to individual patients in a particular cultural setting. Therefore, genetic counselors always have to balance their global and local orientations or the "ethical" and the "moral" (in each of which a further tension between the global-abstract and local-concrete may exist). Moreover, when professionals from two different countries agree about an abstract ethical principle, it does not tell us much about their more culturally-specific moral practices (Wertz and Fletcher, 2004). Let me give a simple example based on my interview materials and textual analysis to demonstrate this claim: both German and Israeli counselors would agree that the general ethical principle of non-maleficence (the obligation to minimize harm) reflects one of their main duties. But if we look deeper, we will find that the meaning of "doing no harm" in the context of prenatal genetics is quite different in the two cultures: Israelis stress doing no harm to the mother (and her family), whereas Germans understand this demand as doing no harm also to the fetus, or to society in general. The results of the regressions clearly show that the sharply differing advice concerning the use of genetic technology given by counselors in Israel and Germany – who share the same systematized knowledge typical of professions (Parsons, 1954) whose global reach can be demonstrated by the existence of international ethical guidelines in the field of medical genetics, (see: Proposed International Guidelines on Ethical Issues in Medical Genetics and Genetic Services. WHO, 1998), as well as by the principle of non-directive counseling (which aims at displaying sensitivity to the personal and cultural background of consultants) – is strongly affected by what the survey categorizes as "nationality".

In fact, the regression results show that the nationality effect still holds and remains the single most effective factor in predicting PND practices when controlling for the other variables. "Nationality" was found to be a significant factor in determining support for abortion in general (both selective and non-selective), parental autonomy, and the rational planning of life with the aid of PND and in predicting whether disability will be seen as good for society.

In fact, the analysis shows that "nationality" affected all the dependant variables but one: support for eugenic ideas. This can be explained by the fact that typically eugenic ideas are nowadays strongly censured throughout the Western world, as part of what Kleinman (1995) terms "ethical" abstract global knowledge about "the good". However, by shifting to the level of local morality or by expressing similar ideas, though not in a clearly eugenic fashion, cracks in the "right ethical position" are revealed. For example, when asked to remark about a statement saying "*It is socially irresponsible to bring an infant with a serious genetic disorder into the world in an era of prenatal diagnosis*" in 1993–95, 90.7% of the German counselors disagreed, while only 19% of the Israeli counselors rejected that claim. In 2000–01, differences were still very large, as 87.8% of the German counselors disagreed, while only 34.4% of their Israeli counterparts shared this opinion (see Chapter 7). Thus, as claimed by Rapp (1999), PND and its practices are simultaneously broadly international in scope and highly embedded in national particularities.

THE EFFECT OF PLACE OF TRAINING

Place of training was only found to affect support for the principle of parental autonomy, which was stronger among counselors who had studied abroad. This is easily explained, as the majority of the German and Israeli counselors who had trained abroad had done so in the US, where patient autonomy is very highly extolled. The overwhelming cultural belief in autonomy among Americans was reflected in Wertz survey results (Wertz, 1999), which showed that it causes physicians never to refuse a service, even if it provides no known medical benefit. Thus, if in the context of reproductive genetics we understand globality as the effect of American professional culture on counselors in both Israel and Germany, we can see that this influence is limited only to their respect for patients' autonomy and does not affect the other dependent variables. Other independent variables dealing with professional characteristics had a minor effect on counselors' moral practices or had no effect at all.

THE EFFECT OF MEDICAL SPECIALTY

Medical specialty affected the counselors' eugenic ideas in that pediatrician-counselors were found to be significantly more supportive of eugenic ideology and practice. The major difference between them and the rest of their professional group has to do with the fact that pediatricians have much closer contact with disabled children and their families, which could serve as an explanation for their stronger "eugenic" tendencies. Indeed, in the interviews I conducted with counselors in Israel and Germany, a number of pediatricians explained that their strong support of selective abortion resulted from the horrible suffering that they had witnessed in children and their families during their careers. For example, one Israeli pediatrician-counselor said: "*You cannot imagine how sad it is to see a child plead for his death or have his parents ask you not to do everything to keep him alive. Such situations have made me sure that selective abortions are the right choice*". On the other hand, another Israeli pediatrician-counselor who had been raised in Europe, said that being exposed to the heroism of families with sick children and to the meaning it endows upon their lives, made her very cautious about thinking that she can know in advance what the best interests of the parents or the child are. Be that as it may, the survey results clearly show that most pediatricians, in both societies, do not feel the same way. Being an obstetrician, as opposed to belonging to another medical sub-specialty, had no specific effect on counselors' opinions. Non MD counselors are hardly represented in the survey since they did not exist in Germany at the time of the research, and were not asked to participate during the first survey in Israel. Due to these limitations, differences between them and other professional groups could not be attested to within this study.

THE EFFECT OF GENDER

While it is true that men (especially future fathers) often take part in reproductive decision making, PND is always applied to women's bodies, be it through blood tests, ultrasounds, CVS, amniocentesis or, in worse scenarios, selective abortion.

As such, PND represents a deepening of two different tendencies. The one allows women more control over reproduction (especially since the invention of the contraceptive pill), while the other reflects the continuing medicalization of childbirth (Oakley, 1984) and the subsequent struggle between fetal and maternal rights. New genetic reproductive techniques also pose contradictory demands on women: they are expected to love their child and take care of it from the moment of conception but must also be willing to abort it if a genetic problem is detected (Katz-Rothman, 1986). The tests also reflect a gendered attribution of responsibility for family health to women. Feminist writers such as Asch and Geller (1996) have suggested that because feminism has long struggled with the relationship between the biological fact of one's sex and the social construction of gender, feminist theory can serve as a model in the struggle to understand the relationship between the biological fact of one's genes and the social construction of inheritance, disease and disability, and thus for fighting genetic discrimination. However, international surveys of geneticists have not found a universal "women's ethic". On the contrary, Wertz's survey results showed that culture and professional locus affected ethical views much more than gender (1997).

My own findings, however, show that gender does have an effect on attitudes towards abortion and the rational-scientific planning of life. Firstly, and as expected, women were found to be more supportive of their own right to abort. Secondly, being a woman affected attitudes about the rational-scientific planning of life with the aid of PND, as women counselors were found to be less inclined to rely on "the verdict of science" in reproductive decisions. Writing about gender and science, Fox-Keller (1985) argues that rationality and objectivity are ideals that have a long history of identification with masculinity, which led to the belief that women cannot, or should not, be scientists. While women are well represented in the field of human genetics, my findings suggest that women scientists' readiness to lean on science in order to make reproductive decisions is more limited than that of men. Thus, women seem to allow themselves more ambivalence regarding their own professional knowledge and its ability to lead to "doing the right thing", as they may not be confined to a rational-scientific way of thinking. It may also be that they are more attentive to their embodied experience. While I cannot argue for a fundamentally different "women's ethics" concerning the use of reproductive technologies, women's attitudes towards their own right to abort, as well as towards the rational-scientific planning of life with the aid of PND, were clearly found to be different from those of their male counterparts.

THE EFFECT OF RELIGIOSITY

Previous research has shown that the scientific genetic discourse is peppered with aspects of the imagery, vocabulary and even conceptual concerns that were articulated hundreds of years ago by Western Christianity and is not a "purely" scientific, secular discourse (Rabinow, 1996). This raises the question of whether the level of religiosity of the scientists in this study bears on their moral practices. My findings concerning genetic professionals do not show religiosity to be an important factor underlying their moral practices, as it was only found to affect attitudes towards non-selective abortion. In fact, the survey results show that in both countries, the more religious the counselors were, the more likely they were to oppose abortion in general. This finding is quite surprising, because while Catholicism strongly condemns abortion and attributes fetuses with the full rights to "life" from the very moment of conception, Judaism is far less critical of it. This is due to the fact that in Jewish doctrine the mother's rights prevail over those of the fetus and the whole Catholic notion of the entry of the soul is quite foreign to Jewish thought (Jakobovits, 1967).

Since being Jewish and being Israeli completely overlap in my sample, in an analysis not shown here, I used the same regression models described above, in order to check for the net effect of being Catholic (versus Protestant or not religiously affiliated) in the German part of the sample. Despite differences between Catholic and Protestant views on abortion, it was once again religiosity and not religious affiliation that was found to affect German counselors' attitudes. This finding is in line with that of Legge (1983), who discovered that in Germany it is level of religiosity rather than denomination that best predicts pro-life attitudes.

THE EFFECT OF TIME

As my study builds on two samples taken approximately seven years apart, this paper also looks at the differences between the two time periods. Genetic knowledge is known to develop rapidly and ethical attitudes concerning genetics are constantly shifting, in a reciprocal process. At the same time, the genetic counseling profession itself is growing very rapidly (for example, in Israel there were about 30 genetic counselors in the mid-90s, and about 50 in 2001). In line with a long tradition in the social sciences that critiques the alleged benefits of technological progress,[4] Ulrich Beck (1992) has warned that progress necessarily implies unplanned access to harmful practices and that the latest research results constantly open up possible new applications at a very rapid rate, meaning that the process of implementation is practically uncontrolled and uncontrollable. This makes it especially important to question how time affects moral reasoning about scientific innovations, particularly in the field of genetics, which is controversial, new and advancing extremely quickly. Therefore, even a gap of seven years may prove to have an effect on ethical trends in the field.

Furthermore, science and technology are often referred to as very powerful social tools, opening up options, which, in an almost deterministic manner, subsequently become social and professional obligations that cannot be rejected. Although "hard" technological determinism – understanding technological change to determine social change in a prescribed manner – has been rigorously criticized (Kline, 2001; Smith and Marx, 1994), genetics is often primed with a coat of dire technological deterministic colours (Timermans and Berg, 2003 p. 101). Accordingly, when it comes to human genetic engineering, well-known and highly esteemed social critics such as Jurgen Habermas (2001), still assume that technical reason cannot be halted.

By looking at the effect of time on the attitudes of genetic counselors, this survey's findings challenge such a line of thought. As a matter of fact, the analysis shows that the second most effective factor influencing genetic counselors' practices is the passing of time. In the second survey (2000–01), counselors significantly toned down their eugenic ideas and their support for the rational-scientific planning of life. In addition, their support of the idea that the disabled are good for society grew significantly stronger. This clearly means that the rapid developments in the field of human genetics have not caused professionals to uncritically "relax" moral norms. On the contrary, the tendency revealed in this study is towards more scientific and professional caution.

Hence, the findings of this longitudinal study prove that scientific knowledge does not always work in a foreseeable fashion, at least with regard to moral practices. However, it is hard to tell whether counselors' cautiousness is the result of a superficial acceptance of ethical criticism or a genuine shift in attitudes and conduct. Because I have examined two quite close periods of time, further longitudinal studies are needed in order to confirm and further explain the discovered tendency.

Age and years of professional experience had no effect on the counselors' opinions. Being older means belonging to a generation that lived through the Second World War or having experienced life shortly after the collapse of the Nazi regime. However, temporal proximity to the eugenic atrocities of Germany's past did not affect counselors' positions in either of the countries. Being new on the job might cause people to behave more in accordance with the prescribed norms and to feel

[4] See, for example, the various writings of members of the Frankfurt School.

less competent about forming their own practices (or *vice versa*), but nor was this assumption proved in this study, as no differences were found between people with little or much professional experience.

CONCLUSION

The major finding presented in this chapter is that German genetic counselors practice tolerant counseling, which allows for more children with genetic irregularities to enter the community, while Israeli counselors practice far more pro-abortion counseling. This is despite the fact that they all share the same systematized global knowledge typical of professions and in spite of the nondirective ethos of genetic counseling. Indeed, the regression results show that while cultural, professional and personal dependent variables had either a minor effect on counselors' moral practices or no effect at all, the "nationality effect" was found to be the single largest predictor of genetic counselors' practices in Israel and Germany when controlling for other variables. Specifically, Israeli counselors were found to be significantly more pro-abortion, in favor of parental autonomy and supportive of the rational-scientific planning of life with the aid of PND but less likely to concur with the idea that the disabled are good for society. Thus, as opposed to the idea that professionals are globally oriented or that they follow their profession's ethos of value neutrality, I argue that what determines professional permissiveness towards genetic prenatal selection is located within national-cultural meaning systems, more than within global professional ethics.

Quite surprisingly, the second most effective factor influencing counselors' practices was the passing of time. Contradicting theories of technological determinism, which understand scientific knowledge to predetermine its application, it was found that rapid developments in the field of human genetics did not cause the professionals to "relax" moral norms in an uncritical manner. Rather, the passing of time caused the professionals in this study to become more cautious regarding their practice. However, more longitudinal studies are needed in order to confirm this tendency.

Having demonstrated that constant and stable differences exist between German and Israeli counselors, which are hardly masked or enhanced by other factors but "nationality", the black box categorized in the survey as "nationality" has to be opened up. And so, it remains to be studied what it is about the two national cultures that accounts for such diversity. This chapter's panoramic quantitative findings serve as the incentive for focusing the rest of this study on local meaning systems, which genetic counselors have at their disposal while having to reach their moral practices, focusing on the concept of "life". The next chapter will thus zoom in and provide an in-depth analysis of the issue of selective abortions in the two countries, while showing how different cultural understandings of the status of mothers and fetuses and of the notion of "life", affect the implementation of prenatal genetics in both countries.

CHAPTER 5

ABORTIONS ON EMBRYOPATHIC GROUNDS: POLICY AND PRACTICE IN ISRAEL AND GERMANY

A pediatrician and genetic counselor working in a large city in Germany:

"I had a patient who went through an amniocentesis but did not come to pick up the results on time since she felt safe and well and went on a holiday. When she got home, in the 24th week of pregnancy, she found out that the child had Down Syndrome. She wanted to terminate the pregnancy but could not find a hospital in the area that was willing to perform the procedure. Finally, she decided that maybe this is how it should be and gave birth to the child".

A genetic counselor discussing a medical case at the Israeli genetic counselors' clinical rounds, Winter 2001:

"Last month an Ashkenazi [Jews of European Origin, YHD] couple with two healthy kids at home came to our clinic. They were quite an educated couple, with a grandmother who works as a nurse in the hospital. They reached us very late in the pregnancy with problems that it took us a while to understand. During the 30th week of pregnancy an ultrasound examination showed some obstructions in the intestines. With the passing weeks, the symptoms became worse. We suspected cystic fibrosis, and did a quick test, which proved our suspicion. In the 38th week of pregnancy the woman arrived at the hospital with contractions but the doctors succeeded in terminating this pregnancy with a stillbirth".

In the two medical situations described above, the medical procedure of "late abortion" is legally permitted (although for different reasons) up to the very end of pregnancy. However, the German reasons for refusing an abortion of this type would most probably not be cited in Israel. There is very little chance that a pregnant Israeli woman carrying a fetus with Down's syndrome would be faced with any professional opposition to her request for an abortion at any stage of pregnancy. Likewise, the Israeli case of a very late abortion or of feticide during birth on the grounds of CF, would most likely not take place in Germany. This is not only due to the presence of contractions, which indicate active birth and place the event itself in a gray area between pregnancy and delivery, but is also related to attitudes in the German professional sphere towards abortions in general, selective and "late" selective abortions in particular, and the separate issue of cystic fibrosis patients. However the same situation, when related to an audience of Israeli genetic counselors during clinical rounds, triggered no apparent moral discomfort. Although the audience noted that earlier detection would have been better, as it could prevent the unpleasant procedure of "an abortion during birth", on the whole it was viewed as a medical success story. This chapter aims to explore these remarkable differences. In a broader sense, the chapter will contribute to our growing understanding of the politics of the beginning of "life", and of the meaning of the bio/cultural concept of "life", within the nations under study.

REPRODUCTIVE GENETICS AND THE ABORTION DEBATE

The abortion debate and the legal compromises surrounding it, have received much political attention in Western democracies in the past 40 years, as it is a major arena for controversies that expose tensions at the heart of liberalism, feminism, family ethics and respect for future human life. Generally speaking, abortion laws, which determine what children are accepted into society, are selective to begin with, in that they sort out socially unacceptable mothers and children from socially desirable ones. They are also restrictive because abortion remains a criminal offense unless it is authorized under the conditions set down by law (Amir, forthcoming).

Any discussion or comparison of the uses of genetic knowledge during pregnancy in Israel and Germany, or examination of the cultural notions of "viability" and "worthy" versus "unworthy" lives in the two societies is unavoidably haunted by the specter of the morality of abortion, and specifically, the morality of "late"[1] abortions. Selective and "late" selective abortions are closely related to reproductive genetics, since prenatal genetic diagnosis goes hand-in-hand with abortion to avoid "birth defects", given that treatments and cures lag behind the ability to detect diseases or increased susceptibility to disease in-uterus.[2] The policy and practice of selective abortions in Israel and Germany, a procedure which determines what children are accepted into society on the grounds of their expected health, can best be seen through analysis of "late" selective abortions. There are a number of reasons for this. Firstly, abortions on embryopathic grounds are no longer registered in Germany following a modification of the 1995 abortion law (discussed below) and thus, they do not officially exist. Consequently, the only way to try and count selective abortions in Germany is via late abortions, which are mostly selective, or done on the grounds of an embryopathic indication. Otherwise, if the woman's health is threatened but the baby is wanted, these "late" abortions would tend to be termed and registered as premature births, since fetuses are now capable of survival at an early age – even before the 20th week in rare cases, (German National Ethics Council, 2003) and of survival without major long-term damage after the 25th week of pregnancy. Second, "late" abortions – those performed after the 22nd week of pregnancy, from which point the child is "viable" (able to survive outside the womb) – are far more problematic ethically than earlier abortions, as

---

[1] The term "late" abortion is a common medical euphemism for "early birth".

[2] In Germany, genetic counselor's guidelines state very clearly that prenatal diagnosis may be done in order to help parents prepare for the birth of a child with special needs. And yet, in the case of a positive diagnosis, for example DS, most pregnancies are terminated. In Israel preparing for birth after a positive diagnosis is not often mentioned as a goal of counseling. When an Israeli woman or couple receiving counseling decides to keep a "problematic pregnancy", they do not get any special support. On the contrary, Israeli counselors have admitted being disappointed if this type of pregnancy is not terminated. When I asked whether special help is offered to parents who wish to keep such a pregnancy, Prof. Zlotogora, the head of the department of Community Genetics in Israel's Health Ministry, replied that most of these people are religious, their beliefs in their decision and convictions are very strong and therefore they need no special care. The ones who need support in his view are those who decide to terminate a pregnancy and have to face the difficulties of such a decision.

they entail feticide preceding the abortion procedure. Because they are such an ethical borderline case, "late" abortions trigger extra moral concern and shed light on the tensions surrounding the issue of abortions. Third, due to the current state of medical knowledge, medical problems in the fetus are quite often detected late in pregnancy, making the issue of late selective abortions for 'defects' an issue in itself.

LAW ON PAPER VERSUS LAW IN PRACTICE CONCERNING "LATE" ABORTIONS ON EMBRYOPATHIC GROUNDS IN ISRAEL AND GERMANY

In most European countries, "late" abortions are not legal: most countries permit abortions until the 12th week of pregnancy, or until the 22nd week when an abnormality is proven to exist (Harris and Reid, 1997). "Late" abortions are permitted in rare cases, such as in Germany (where non-medical abortions are forbidden after the 12th week), and Israel, which has no time limit whatsoever for abortions.

Germany: in 1995, the explicit embryopathic indication for an abortion was abolished from the law in Germany (Pregnant Women's and Families' Aid Amendment Act, 1995). The campaign to change the law was spearheaded by the Catholic Church and the disability lobby, both of whom oppose any worldview that sees "abnormal" fetuses as having a reduced right to life. According to Schlegel (1997), this elimination of the embryopathic clause is also related to German history, in that it is a belated and oblique condemnation of the eugenic terror of the "Third Reich".[3] However, in practice, this symbolic-declarative change in the law did not close the door on "therapeutic" abortions, which had previously been permitted only until the 22nd week of pregnancy. On the contrary, along with the abolishment of the embryopathic indication, the new law now allowed for abortion at any stage of pregnancy, if the mother's physical or mental health were deemed to be at risk.

The German law reads:

> *"...the termination is allowed, if from a medical point of view, it prevents danger of life or danger of severe impairment of the physical or psychological health of the pregnant women in the present or in the future and this danger cannot be abolished by other means to be offered to her"*
> (Abortion law 1995, paragraph 218a, part 2. Translated in Wegner and Becker,1997)

Furthermore, since 1995 the requirement for counseling, the three-day interval between counseling and termination and the separate statistical recording of these terminations have ceased to apply (German National Ethics Council, 2003). Thus,

---

[3] Public involvement in these issues differed in the former East and West Germany. Public discourse in the former West Germany almost always included references to selective abortion practices and medical experiments conducted by the Nazis. Such references were nearly non-existent in former East Germany. Erikson suggested that this was a result of West German post-war educational and political policies, which acknowledged guilt for the Holocaust (Erikson, 2001). Conversely, East German post-war narratives emphasized a redemptive, exculpatory Soviet victory over fascism, which exempted East-Germans from responsibility for the Holocaust. (Herf, 1997, Welsh, Pickel and Rosenberg, 1997).

paradoxically, since 1995 "late" abortions are actually permitted for the first time in German history. Since selective abortions are resituated within the mother's medical clause (which has no time limit), "late" selective abortions can be performed using the mother's claim that she is physically or mentally endangered by the pregnancy. In the words of the German National Ethics Council: "It is recognized (and this recognition reflects the will of the legislator) that the medical indication also includes the "substitute indication" for the cases that were formerly embryopathically indicated" (German National Ethics Council, 2003 p. 63). Thus, the explicit abolishment of the eugenic indication went hand-in-hand with the implicit permission to abort for this same reason until the end of pregnancy through the back door of the mother's health and well-being. In consequence, by rendering abortion dependent upon a medical condition in the mother, selective abortions were released from any time constraints and protection for the fetus was scaled down. According to the new legal formulation, a pregnant woman wishing to abort because of an explicit fetal condition was now forced to justify her request for an abortion on the grounds that she would be unable to deal with an "abnormal" child. Nonetheless, although the words "eugenic", "medical" and "embryopathic" are no longer mentioned in the German abortion law, everybody uses them and everybody knows what they mean. In fact, a major discrepancy now exists between the rhetoric of the formal law, which eliminated the embryopathic clause and its actual potential applications, which allow for late selective abortions.

Israel: The Israeli law (Penal Law, 1977. Interruption of Pregnancy 312–321) acknowledges an embryopathic indication as a just cause for abortion throughout pregnancy and states that termination is allowed in case "the newborn is likely to have a mental or physical defect". As a result, an Israeli woman wanting to abort for an embryopathic reason is not made to feel that she must put her own motherhood in the balance. However, in 1995, medical committees known as Abortion High Committees were established in Israel in order to make decisions about "late" terminations, thereby re-emphasizing the power of medical practitioners as legal gatekeepers of abortions (Petersen, 1999, 2000). A number of professionals sit on these committees, including the director of the hospital in which the abortion is to be performed, the head of the genetic institute, the head of the gynecological ward, the head of the neonatology ward and a senior social worker, or their substitutes.[4] At the end of 2004, thirteen such committees operated in Israeli hospitals with licenses from the Ministry of Health.

Despite the fact that the rhetoric of the laws is different and in itself reflects cultural differences, whereby Israeli law explicitly allows for abortions on embryopathic grounds while German law does not, both practically permit selective and "late" abortions, either explicitly (the 'Israeli case) or implicitly, through the 'mother's health clause (the German case). Still, by only examining the laws as they exist on the statute books and not looking at actual practice, it would be easy to

[4] For the statute (in Hebrew) see: www.health.gov.il/downlad/forms/a2346_r70c_93.pdf

draw the erroneous conclusion that the German law's recent change was merely an example of lip service, admitting selective abortions through the back door while putting the blame on the mother. Indeed this is partly true, since after the law was changed in 1995 there was a dramatic rise in the number of late abortions performed in Germany (most likely on embryopathic grounds): from 26 in 1995 to 157 in 1996 (Federal Statistics Office, Germany). However, the statistical data from which I depart in making my argument suggests that even though "late" abortions have been legal in both Israel and Germany since 1995, comparatively, many more "late" abortions are performed in Israel than in Germany.

THE PRACTICE OF "LATE" SELECTIVE ABORTIONS IN ISRAEL AND GERMANY

Israel: In contrast to the stipulation in the amended German 1995 law, there is no registration problem in Israel, since the law acknowledges the embryopathic indication as a just cause for abortion throughout pregnancy. Although the Israeli law leaves the door open for "late" terminations for any statutory reason, in practice the vast majority of "late" abortions are approved on medical grounds. According to the Israeli Health Ministry (www.health.gov.il), in the years 1995–2002 over 80% of "late" abortions approved in Israel were justified by the clause in the abortion law that allows for termination if the fetus is suspected of having a physical or mental malformation.

Having a "late" abortion on the grounds of the fetus' medical condition is easy in Israel. Therefore, in the interviews I conducted, the counselors were often anxious about the fact my study could unleash a societal debate on abortion and in particular "late" abortion policy, rather than 'letting sleeping dogs lie'. Similarly, Cohen-Almagor and Snir (2000), who studied the influences on the decision-making process in the Abortion High Committees, were faced with reluctance on the part of some potential participants to ignite a delicate social issue that is currently non-controversial. Nonetheless, their major findings show that the high Israeli abortion committees are very permissive with "late" selective abortions.

Germany: despite the fact that abortions for medical reasons are permitted right until the very end of pregnancy, this procedure is far less common in Germany than in Israel. In fact, "late" abortions have elicited considerable opposition in Germany. For example, after the embryopathic clause was abolished, the BÄK ("*Bundesärztekammer*", the German medical professional association, which is an important professional gatekeeper) declared that it rejected "late" abortions (*Bekanntmachungen der Bundesärztekammer*, 1998). As a rule, they claimed, viability should be regarded as setting the time limit for abortion, except in those rare and limited cases in which feticide before abortion is justified (for example when the baby will not survive long after birth).

Table 1 presents the percentages of "late" abortions as a proportion of the total number of abortions in Israel and Germany from 1996 to 2002.

TABLE 1. Percentage of "late" abortions (23rd week and later) as a proportion of the total number of abortions in Israel and Germany in the years 1996–2002

	Total no. of abortions	Total no. of "late" abortions	% of "late" abortions as a proportion of the total no. of abortions
Germany	924,014	1207	0.13%
Israel	132,907	1400*	1.05%

Sources: Federal Statistical Office Germany (http://www.gbe-bund.de). Israeli Health Ministry (www.health.gov.il).

* The Israeli Health Ministry does not provide exact numbers of "late" abortions. According to its publications, in the years 1995–2002, on average, 200 "late" abortions were approved each year.

In Israel, "late" abortions constitute 1.05% of the total number of abortions, whereas in Germany they constitute only 0.13%. Thus, the rate of "late" abortions in Israel is 8.07 times higher than in Germany.[5] This difference remains considerable, even if we take into account the fact that due to public opposition to "late" abortions in Germany, some abortions are performed without being properly registered or that a minority of "late" abortions in Israel are performed for non-embryopathic indications.

THE ROLE OF PROFESSIONALS IN ABORTING ON EMBRYOPATHIC GROUNDS

A socio-legal approach to the law suggests that "gaps" between the law on paper and in practice are a typical feature of legal practice (Lee, 2003), as the law is never complete and comprehensive but rather, its enactment is affected by social dynamics (Galligan, 1995). In the following I will present a comparison of two such social dynamics, in both of which medical practitioners are given the mandate of gatekeepers to abortion services and thus implement the law in practice (Petersen, 1999).

In Israel, genetic counselors do not have the sole power to decide whether an abortion should be permitted or not, though in practice their recommendations are rarely rejected. The procedure is as follows: whenever an embryopathic problem is detected during pregnancy (by a gynecologist or genetic counselor), it is most often genetic counselors who are consulted. After meeting with the potential mother or parents and only with their consent, the counselors write a recommendation to the abortion medical committee (either the regular or high committee, depending on the stage of pregnancy), which is the body that ultimately approves or denies requests

[5] Michael Gross (1999) compared the rate of late-term abortions as an abstract of the annual number of abortions in Denmark, the US, the UK, Canada and Israel. His findings suggest that the Israeli rate is five to ten times higher than the rates reported in the other four countries. Gross explains this to be due to different norms for fetal personhood.

for legal abortions. Amir and Biniamin (1992) studied the practices of regular abortion committees in Israel (composed of a social worker and two physicians, one of whom must be an obstetrician/gynecologist and one of whom must be a woman, usually the social worker). They found a complex mechanism of social control, in which the abortion candidate is symbolically punished for her "bad behavior;" i.e., having unprotected sex and for wishing to abort. However, despite this symbolic punishment, according to Amir and Biniamin the actual refusal rate of the Israeli committees is very low (about 5%). Furthermore, in contrast to the situation in Germany, Israeli committees are not required by law to adopt a pro-life position or to provide any counseling at all and any feelings of guilt they may trigger among women are not perceived as related to the fetus' right to life, but rather to the woman's supposed promiscuity. All in all, it is likely that a woman who wants to have an abortion in Israel will face a liberal committee that will assist her in doing so. The same is true in the case of "late" abortions, which have to be ratified by High Committees and had a refusal rate of less than 10% from 2000 to 2001 (www.health.gov.il). Since these committees operate without written regulations, the majority opinion of the committee members is required to reach their decisions. However, Cohen-Almagor and Snir (2000) found that in most cases the committee members reach their decision unanimously, as they share the same cultural background and hold the same ideas about desirable and undesirable children. Thus, this professional-cultural consensus helps them reach delicate decisions that are usually justified on the grounds of respecting the mother's autonomy.[6]

In Germany, as in Israel, genetic counselors can recommend abortion but the final decision is not theirs. Hospitals or gynecologists may refuse to perform the termination and reject counselors' recommendations. My interview materials suggest that refusals, especially for "late" abortions, are not rare. Moreover, there have been cases where pro-life groups have threatened hospitals and doctors because they performed abortions and in some such cases the practice has been stopped.[7] A number of my German interviewees said that they preferred to steer clear of "late" abortions because they feared losing their credibility or being exposed to pressure from their peers or the media. In both Germany and in Israel physicians have the legal right to refuse to perform an abortion if they find it goes against their conscience (*Berufsordnung für die deutschen Ärztinnen und Ärzte*, 1997 and section 318 of Israeli criminal law).

However, in contrast to Germany, this section of the Israeli law was never mentioned by any of the Israeli interviewees, who never cited a situation in which

---

[6] The Israeli law (like the German one) does not extol women's autonomy, since the abortion committees are supposed to regulate their freedom (although they rarely do so in practice). However, when it comes to abortions on embryopathic grounds, autonomy becomes a highly respected value, which obviously serves as a way to avoid moral conflict for professionals, who can thus place the moral decision entirely in the mother's hands.

[7] Personal communication, interview materials. The hospital referred to is the University Hospital in Essen.

Israeli gynecologists had refused to perform an abortion. Thus, my data suggests that despite the similar legal options existing in Germany and Israel, the implementation of the law by professionals is very different in both nations.

Nonetheless, the total responsibility for abortion practices cannot be placed entirely on the shoulders of the professionals: the wording of the law obviously also has implications. Analyzing the language of the Israeli law, Shapira (1995) writes:

"The phraseology adopted by the Israeli legislature appears to reveal an unmistakably lenient policy on abortion for eugenic reasons. The physical or mental defect justifying interruption of pregnancy need not necessarily be extensive or grievous. An ordinary, perhaps even a relatively minor, defect may suffice. Furthermore, the defect need only be "likely" as distinct from certain or probable. The term "likely" seems to denote a more reasonable possibility, which may fall short of a near certainty or a high probability" (Shapira, 1995, pp. 24–25).

In contrast, as explained above, in Germany the language of the law does not allow for eugenic abortions (which is a strong symbolic statement in itself) but only for "late" abortions on the grounds of grave danger to the mother's mental or physical wellbeing. Yet, despite these differences in rhetoric, it is still professionals in both countries who are called upon to interpret what justifies a "late" abortion. Moreover, the language of both laws is general enough to leave considerable room for interpretation because it does not give clear guidelines covering the full range of possible medical scenarios. Hence, in both Germany and Israel, it is professionals who decide what might cause a "severe impairment of the physical or psychological health of the pregnant women in the present or in the future" (in Germany), or whether there is any likelihood of a "defect" (in Israel). In this respect, professionals' decisions obviously entail considerations beyond purely medical or legal stipulations, as there is no accepted professional definition of what constitutes a defect, nor where to set the yardstick defining whether an impairment is severe or mild, or what is a high or low likelihood for it to occur. Both German and Israeli abortion laws appoint medical practitioners as legal gatekeepers to abortion and "late" abortion services in order to mediate and fill the "gap" between the law and its practice and thus to interpret and implement the law (Petersen, 1999). In fact, in the case of abortions in both Israel and Germany, doctors, representing the medical establishment with its power as an instrument of social control and as a self-regulating profession with the right to clinical autonomy, rather than pregnant women or the legal system (the formal legal system of lawyers, courts and judges, is not actually involved in most abortions), decide whether an abortion is permissible, even when the reasons are not medical (Petersen, 1999, 2000). This role is akin to that played by other professionals in modern societies, who set moral boundaries demarcating good and evil, deviance and normality, insiders and outsiders and thus control values, beliefs and their related social practices (Abott, 1988; Davis, 1985; Kraus, 1996). Yet, professionals may place strong *de facto* restrictions on the options offered by the law (as in Germany) or alternatively they may make it very accessible (as in Israel).

However, by stressing the importance of professionals as gatekeepers to abortion, I am not suggesting that the legal system itself, or the specific abortion laws, do not matter. Rather, this chapter presents a unique case, since although major differences are registered in the rhetoric' of the laws at hand, in practice they are very similar. Thus, I argue that what mainly accounts for the different enactment of the laws are professionals' interpretations. Let us now study the counselors' common understandings concerning the status of mothers and fetuses and see how they assist them in filling their roles as gatekeepers to abortion services.

GENETIC COUNSELORS IN ISRAEL AND GERMANY: DIVERSE VIEWS ON THE ABORTION DEBATE

When asked to comment about free (non-selective) abortions before viability, Israeli counselors were more supportive of abortions on demand than German professionals. In the questionnaire, for the statement: "*A woman should have an unqualified right to abortion before viability*", the majority of Israeli counselors agreed (57.1% and 59.3%), whereas the majority of German counselors disagreed (61.5% and 59.6%). (See Table 2)

Likewise, during interviews, in response to the question "What is your opinion on abortions and the moral standing of the fetus?[8]" Israeli interviewees manifested little inner conflict (in contrast to the German interviewees). For instance, a female genetic counselor with a PhD who works in one of Israel's largest hospital genetic institutes said:

"I have nothing against abortions. I think they can reduce suffering. In any case I respect the parents' wishes and if they want to abort, I see it as their full right. Generally I trust the parents that when it's a wanted pregnancy, as is usually the case when people reach me; they will not rush into abortion for stupid reasons. And it is also not up to me to decide what a major or a minor problem is. Abortion is very hard for women, it involves a lot of grief, and I sympathize with the mother's pain. But the fetus? It is really nothing to me. Only a potential for life, with no rights."

TABLE 2. Support for Abortion on Demand

	1994–96					**2000–01**				
	Strongly Agree + Agree	*Neither Agree nor Disagree*	*Disagree + Strongly Disagree*			*Strongly Agree + Agree*	*Neither Agree nor Disagree*	*Disagree + Strongly Disagree*		
Germany	*23.6%*	*14.8%*	*61.5%*	*100%*	*N* = 185	*26.3%*	*14%*	*59.6%*	*100%*	*N* = 57
Israel	*57.1%*	*19%*	*23.8%*	*100%*	*N* = 22	*59.3%*	*25.9%*	*14.8%*	*100%*	*N* = 31

[8] The Hebrew language does not differentiate between an embryo and a fetus. In Germany I used the term "fetus". However, German counselors used the terms fetus, embryo and even "pre-embryo" (a controversial term in itself) in their responses to questions about their attitudes.

Similarly, a pediatrician and genetic counselor who heads one of Israel's genetic hospital units did not perceive the fetus as having autonomous rights or as independent from its mother and thus, did not see abortions as a major moral problem, early or late in the pregnancy:

"The fetus belongs to the mother. It is her business and she is allowed to do with it things that I might not agree with. I am always there first of all for her. After all, she is the one who would have to raise this child. For me, the fetus is part of the woman's body, until very late. Of course, in the 30th week of pregnancy it is harder to see it that way but I would hardly see the fetus as independent".

Another female counselor who works in one of Israel's major genetic hospital units was very straightforward about the connection between prenatal genetic diagnosis and abortions:

"The goal of genetic counseling is to prevent the birth of children with birth defects, if that's what the parents choose to do. How do we do it? By abortions. Right from the start our role is to lead people to abortions. It might sound cruel but this is our job, by definition. If I objected to abortions or cared too much about the fetus, I could not do this job. And anyway, I think it is better for the fetus not to be born with a severe disease".

In contrast, German interviewees had a completely different view of the fetus, its status, its rights, its relationship with the mother and their job definition. For example, a female genetic counselor and gynecologist who works in the private sector answered as follows:

"It's hard for me to say what the fetus is. But basically I think life begins upon implantation [after 14 days, Y.H.D], because before that the baby has not been accepted. But even in IVF treatment, before implantation there is potential for life. After all, those early cells can be implanted. That's why they should be treated very carefully, like the hope of a baby... There is no real point in time where life begins, that before it the embryo is just stem cells or chromosomes..."

Similarly, a male doctor and clinical geneticist who heads a large genetic institute spoke of the fetus in terms of a "life", and of "late" abortions in terms of "feticide". He said:

"Life begins early. But still there is a difference between 8 stem cells and a human-shaped embryo and between prenatal and postnatal life. I think that late abortions are feticide. But if you know that the child will die anyway soon after birth, for example with trisomy 13 or 18, I don't oppose late abortions. But if it is only, for example, a kidney problem, I think abortion at late stages is feticide.

Likewise, a female doctor and human geneticist, who works in a hospital, placed the 14-day old fetus' rights on a par with other family members:

"I can live with a stage model. After fertilization and before implantation [first 14 days, Y.H.D], I don't think it is a highly protected human being. It's only a pre-embryo and therefore I do not oppose stem cell research. But after that, the embryo must be protected. Within the family context it has a right, like anyone else. Not more or less but like the others. Late abortions after viability, in which the embryo has to be killed before it is aborted, are something I oppose".

Thus the German counselors who were interviewed and whose profession is closely related with selective abortions were far more ambivalent about this procedure than their Israeli counterparts. Whereas most Israeli interviewees repeatedly emphasized the supremacy of the mother's rights and wishes over the fetus' non-rights, German counselors spoke of the fetus in terms of a "life" with autonomous rights. Furthermore, all but one of the Israeli counselors I interviewed saw no major moral problem with "late" selective abortions. In contrast, their German counterparts almost unanimously expressed antipathy or at least serious discomfort regarding the issue, even terming "late" abortions "feticide".

When asked specifically about their opinion on abortion laws, German interviewees were very often critical about the legal underpinnings of their work and felt that the legislation authorizing "late" abortions was something they morally opposed. They thus felt justified in not taking full advantage of the abortion options provided by the law. For example, a female genetic counselor and gynecologist who works in the German private sector said:

"Late abortions are a big problem for me because of the feticide involved. I am not willing to perform them as a doctor. I really preferred the former law with the 24 week limitation. Viability is really the limit for me".

Other German counselors felt that the attitude of the German legal system towards the fetus was inconsistent and confusing in that the enactment of the Embryo Protection Law (EPL) contradicts the abortion law. For example a German (male) head of a hospital genetic institute, said:

"We in Germany have a big problem. On the one hand it is possible to abort until the end of pregnancy but on the other, if you deal with stem cells you may end up in jail. The abortion law is extremely liberal, whereas the embryo protection law is extremely restrictive. The two laws don't make sense together".

In sharp contrast to this counselor's ambivalence, all Israeli counselors I interviewed felt that the Israeli abortion law was reasonable. Moreover, what seemed to worry some of them was that the law could restrict certain choices. For example, a geneticist and pediatrician who heads one of Israel's largest hospital-based genetic institutes, pointed out to me that my research topic was a delicate matter that was not in "our interest" to evoke. He said:

"I hope you are aware of the fact that there are groups in the Israeli public, like the ultra-Orthodox, who are just waiting for the opportunity to restrict the abortion law. You, as a woman, should be very careful not to collaborate with these forces, which not only work against medical progress but also against feminist interests"

To conclude, as we have seen both in Germany and in Israel, genetic counselors are constantly confronted with moral dilemmas resulting from the Gordian knot that has been tied between prenatal diagnosis and the practice of selective abortion. However, the ethical considerations that counselors in each society make when dealing with these dilemmas differ significantly. German counselors placed the fetus at the center of their ethical considerations and referred to it in terms of a "life", or as an autonomous being, deserving rights and protection. Their Israeli counterparts, in contrast, emphasized the precedence of the mother's rights and wishes over those of the fetus. In fact the fetus was not at all considered by Israeli counselors as an autonomous being with distinct rights. Rather, it was discussed as an integral part of its family, specifically its mother. Therefore, in Germany professionals side with the explicit-symbolic position of the law and restrict the implicit actual possibilities for abortion that it affords, whereas in Israel professionals facilitate abortions. I have not yet elaborated on the context of the professionals' attitude and practice formation. The following section expands on this and relates to the framing of the abortion debate in the two nations at hand. I thus argue that whereas in Israel, history, politics and (counter-intuitively) even religion, act as push factors for selective abortions, in Germany the same forces act against them.

FRAMING THE ABORTION DEBATE IN ISRAEL AND GERMANY: HISTORICAL, RELIGIOUS, LEGAL AND POLITICAL CONTEXTS

Having demonstrated that the professionals' views of the fetus are neither predetermined by the law nor based purely on medical-professional knowledge, the question that remains to be answered is what shapes the professional's legal-medical interpretations? Schoen and Rein (1994) define framing[9] as the underlying structures of belief, perception and appreciation upon which policy positions rest. According to them, the frames that underlie controversy are usually tacit, which means they are exempt from conscious attention and reasoning, as they belong to the taken-for-granted world of policy making. Medical practitioners in both countries are educated scientists who share updated scientific knowledge about the different stages of pregnancy and about embryonic development and pathology. However, as suggested by the empirical data presented here, this does not mean that the two groups hold similar positions regarding the fetus, its rights or its status (as a "life" or not). In the following I will argue that it is the overall framing of the abortion debate in their respective societies which shapes the way in which professionals interpret the "facts of life". However, by saying that, I am not suggesting that the abortion laws themselves are untouched by the framing of the abortion controversy. Rather, despite the practical similarities between the German and Israeli abortions laws, their understanding of the abortion controversy is very different. Let us now understand the historical, political and religious framing of the German and Israeli abortion debate and see how it leads to culturally-specific ways of knowing and acting.

POLITICAL HISTORY OF THE ABORTION LAWS

Germany

In the 1871 penal code of the German Reich, paragraphs 218–220 criminalized most abortions (Grossman, 1997). During the Weimar republic in the 1920s and early 1930s, the first campaigns against abortion took place and the issue quickly

[9] Framing is a social science concept with a very long history starting with the work of Erving Goffmann (1975), which I will not discuss here. (For a more extensive discussion concerning Frame analysis in the context of the abortion debate see: Marx Ferree, Gamson, Gerhards and Rucht, 2002)

became politically explosive (Usborne, 1992). A new abortion law in 1926 contained some liberal changes that were eradicated a few years later by National Socialism (Usborne, 1992). Under the Nazi regime, restrictions on abortion and birth control that had been loosely enforced during the Weimar years were tightened and enforced according to racial and eugenic criteria. Nazi abortion law sharply distinguished between "worthy" and "unworthy" life, forbidding abortion in the former case and demanding it in the latter. Hence, sterilization and abortions (even of advanced pregnancies) on medical and eugenic grounds were legalized and widely performed. By 1943, the incredibly high incidence of illegal abortions during wartime led to the inclusion of the death penalty in abortion law, if an abortion was deemed to have impaired the vitality of the German nation (Grossman, 1997).[10] Contemporary German law and the West German law from the 1970s that it is largely based upon, are in many ways the result of the traumatic disregard for life during the "Third Reich", as post-war Germany could not escape the impact of Nazism on the issue of abortion. For example, in the 1970s a new generation of feminists uncovered the history of the Weimar abortion campaigns (Grossman, 1997) and succeeded in rapidly bringing about a legal change such that from June 1974 to February 1975 a new and less restrictive abortion law was in place in the FRG.[11] This law was similar to the liberal East German abortion law of 1972,[12] although it required counseling by a physician or an approved organization. However, the success of this law was short-lived. A decision of the Federal Constitutional Court dated February 1975,[13] found this procedure to be in conflict with the constitutionally-protected right to life and the State's duty to preserve human dignity. The past was evoked and the State's obligation to see pregnancy carried to term was justified by the memory of Nazi crimes (Grossman, 1997). In the hierarchy of constitutional values, the court held that human life is a central and supreme constitutional value to which women's acknowledged right to self-determination and privacy is subordinate. In addition, the German Constitutional Court chose to emphasize the character of the right to life (of the fetus) as a value of the community rather than as something that belongs to the fetus (Glendon, 1987) and the court also implicitly assumed that human life begins upon implantation. On the whole, during those years, West German law was among the most restrictive in Europe.

During the Cold War, rivalry between East and West Germany provided an opportunity for East Germany to liberalize its abortion law and abortions were far more common under the Communist regime. Accordingly, the East German law of 1972 allowed abortion on demand during the first trimester of pregnancy. This discrepancy between the two German abortion laws became one of the more

[10] Gisela Bock (1991) provocatively argued that the interpretation of National Socialist gender policy as essentially consisting of pronatalism and a cult of motherhood is largely a myth. Grossman (1997) challenges this view.

[11] (1974) BGB1 I 1297–300.

[12] (1972) GB1 of the former GDR I 89

[13] (1975) NJW 573–87 + BVerfGE 39, 1–95.

complex questions to be resolved following unification in October 1990 and was even publicly referred to as "the most emotional issue of unification"(Jarausch, 1994). The main problem was that the West opposed the permissive abortion law of the East. Lengthy negotiations eventually resulted in a compromise to postpone the decision for two years (until 1992), during which time the two laws would continue to apply within their former geographical borders. The most extreme argument in this heated exchange was, once again, an attempt to equate abortion with the genocidal practices of the Nazis (Mushaben, Giles and Lennox, 1997). The first and short-lived solution to the debate was formulated on 25 June 1992,[14] with the German Parliament's adoption of a new law, entitled "The law for the protection of potential life, for the promotion of a pro-child society, for assistance in conflicts concerning abortion and for the regulation of abortion". Its very name captures the story of the political conflicts concerning abortion in unified Germany (Funk, 1993). This new law allowed abortion on demand during the first trimester of pregnancy provided that the woman receives counseling from physicians, psychologists or social workers. At the time it was hoped that this would satisfy the Federal Constitutional Court, which expressed open disapproval of abortion by ensuring that women who wanted abortions would undergo compulsory pro-life social counseling. It also introduced a supplementary package of social security measures to improve the situation of mothers (Schlegel, 1997). However, as in the 1970s, opponents of the law submitted the bill to the Federal Constitutional Court in an attempt to challenge its compatibility with the Basic Law of the State (the German Constitution). In August 1992, the Court suspended application of the new law pending its ruling, which was announced on 28 May 1993.[15] The court based its opinion on three fundamental rights stated in the German federal constitution: the protection of human dignity, the guarantee of free development of personality and the right to life (Schlegel, 1997). In the new decision, women must be informed that their unborn child has its own right to life, and that even though they and their doctors would not be prosecuted for terminations performed in the first trimester, abortion was fundamentally illegal (Grossman, 1997).[16] Hence, abortion in the first trimester (12 weeks) with counseling was rendered illegal but not punishable.

According to Andrea Wuerth (1997), who analyzed the abortion debate in Germany during unification, protection of *all* "life" became the hallmark of post WWII West German morality. During unification, in contrast to the East German State, which was presented as politically, economically and most importantly morally, bankrupt, as it encouraged abortions as a means of family planning, the West (which "rescued" the East) was presented as a moral state that protects and values "life", mothers and the family. Reducing the number of abortions and

[14] (1992) BGB1 I 1398–404.

[15] (1993)NJW 1751–79 + BVerfGE 88, 203–366.

[16] Grossman (1997) explained the lack of aggressive feminist response to the law of 1993 as partly due to the emphasis in current historical scholarship of the anti-rather than pro-natalist aspects of Nazi population policy.

protecting life were espoused as national goals, displacing appeals for woman's rights. Consequently, the "public fetus" came to represent a new class of German citizens, the unborn (Wuerth, 1997). Women who wanted to abort were reminded in the mandatory pro-life counseling that the fetus is a citizen and a member of the national community and seeking an abortion as a result became an act of treason, since women who aborted were seen as abandoning their responsibilities to their national community as its collective womb.

The abortion law went through a third round of legislation known as the "Pregnancy Conflict Law" on 29 June, 1995.[17] This new legislation abolished the eugenic stipulation, leaving only two indications for abortion: medical and criminal. This recent change accounts for the unique formulation of today's abortion law in Germany. There is no thorough explanation of the causes for this last (and not necessarily final) change, which opened the door to "late" selective abortions. The contemporary formulation of the law is replete with contradictions, as it simultaneously respects the positions of groups opposed to selective abortions but also allows such abortions to take place through a different clause of the law, which is less restrictive than was previously the case. Since there are hardly any powerful voices in the German public arguing for the legality of "late" selective abortions, the future of the law remains an open issue.

In conclusion, the complex history of the German abortion law is far more controversial than the Israeli one. Firstly, it is deeply affected by Germany's Nazi past. Secondly, despite feminist efforts since the days of the Weimar Republic to change the law, the dominant attitude in political discourse on abortions in Germany patronizes and marginalizes women and constantly depicts abortions as immoral. Thirdly, in accordance with Catholic doctrines and the ethos that extols the protection of "life" as a hallmark of the new German morality (Wuerth, 1997), German law has repeatedly been restricted on the grounds that it does not offer adequate protection for the fetus, which is considered a "life".

Israel

While conflicts over legislation have periodically surfaced in Israel, abortion is not a controversial issue in Israeli politics and the subject has largely remained tangential to Israeli public debate (Morag-Levine, 1994). Abortion has most often entered Israeli politics in the context of coalition demands made by Orthodox Jewish parties that have conditioned their support for different governments on restrictions on abortion. In general, however, the issue of abortion in Israel seems to elicit more apathy than ambivalence (Morag-Levine, 1994). Moreover, abortion is usually justified or opposed in Israeli culture in terms of arguments about demographic or social distress, and not in terms of feminist or right-to-life discourse. According to Yishai (1993), policy changes in Israeli abortion policy – specifically the liberalization of the law in 1977, which allowed for abortions for social reasons such

---

[17] (1995) BGB1 I 1050–57

as poverty and the restrictions subsequently introduced through the repeal of this broad social clause two years later – reflect two contradictory aspects of Israel's perceived population problem: on the one hand, the link between family size and social distress and on the other, the fertility imbalance between the Jewish and non-Jewish communities. Yishai maintains that the liberalization of abortion was triggered by a growing awareness in the early 1970s of the fact that a large family constitutes a social risk group. This prompted policymakers to legalize abortion and promote a social cause clause. Later on, the repealing of this clause within the framework of coalition agreements was presented as a possible solution to the so-called "demographic time bomb". Both then and now, fertility differentials between Jews and Arabs were seen as large enough to jeopardize the country's Jewish majority. Jews were therefore encouraged to "be fruitful and multiply". However, the restriction of the law had no effect on the number of abortions performed, as women, assisted by liberal committees, framed their abortion requests so that they fit the enabling clauses of the law (Amir and Biniamin, 1992). The fact that the authenticity of women's declarations were never put to test, once again attests to the flexible application of the legislation.

In summary, the history of Israeli abortion policy tells a story of demographic and social concerns and not one of conflict between the rights of mothers and their fetuses. The abortion debate in Israel does not seem to attest to the nation's (im)morality, and protection of the fetus is virtually a non-issue (Gross, 1999). Furthermore, feminists have not made abortion a central theme on their political agenda in Israel (Amir, forthcoming). Likewise, "Efrat", the only Israeli organization fighting abortions, a religious organization which aims at helping women who want to abort because of financial distress, does not reject embryopathic abortions (www.efrat.org.il). A prime reason for this attitude can be found in an examination of the Jewish religion, which does not pose strong opposition to abortion in general or to selective "late" abortions in particular.

RELIGIOUS ISSUES IN GERMANY AND ISRAEL

When discussing the effects of religion in Germany I focus on Catholicism, despite the fact that Catholics make up only 34% of the German population. There are several reasons for this. First, the Protestant position regarding abortion is not univocal (Wildung Harrison, 1998),[18] and in general, abortion is not a major issue on the Protestant agenda. Secondly, Catholic doctrine frames the abortion debate in terms of right to life and hence, serves as the discursive framework for any moral and legal discussion concerning this matter in much of the Christian world. Thus, even opponents of the Catholic view are forced to frame their arguments using terms from Catholic discourse, which proclaims that the fetus is a living human

[18] One can find a spectrum of Protestant positions ranging from Protestant fundamentalists who forbid abortions, to more "old-line" protestants who are far more permissive and to pro-feminist liberationists who support women's right to choose (Wildung Harrison, 1998)

being from the moment of conception.[19] According to Catholic dogma, since the soul distinguishes human beings from other entities and since it enters the fetus at conception, under no circumstances should the fetus be hurt. Furthermore, a newly-conceived human being is on a moral par with adults (including its mother) and a temporal boundary cannot be drawn between an unborn fetus and a newborn baby, making the killing of a fetus murder (Connery, 1977).

The legal discussion of abortion in Germany clearly distinguishes between the mother and fetus, in accordance with this Catholic interpretation. This is best illustrated by the German Constitutional Court decision dated May 1993, which reads:

"The termination of pregnancy must be regarded as fundamentally wrong (unrecht) throughout the entire period of the pregnancy and thereby must be considered illegal. The right to life of the unborn may not be placed, even if just for a limited period of time, in the hands of a free, not legally-bound decision of a third person, even the mother herself" (Quoted in Wuerth, 1997).

The German law also sets a time limit for all abortions, except those which are performed to protect the physical or mental wellbeing of the mother. Furthermore, while abortion committees in Israel do not hold pro-life positions, the mandatory counseling in Germany for women who wish to abort is required by law to strive for the continuation of pregnancy and to inform the woman of the fetus' right to life. Religious views of the relationship between mother and fetus and the status of the fetus then, have a profound impact on the formulation of civil laws.

Only about a quarter of the counselors I interviewed in Germany belonged to the Catholic Church. Nevertheless, the Catholic view of the fetus as a "life" right from conception was strongly reflected in their opinions, even when they declared themselves to be non-religious or non-Catholic. Although the religious undercurrent to the abortion debate in Germany is not usually explicitly mentioned, in the interviews, the counselors repeatedly talked of fetuses using a Catholic frame of reference and described the fetus (even very young ones) as a form of "life" with its own rights.

On the contrary, Judaism does not ban abortion. Ancient Jewish law affords the fetus no legal status, since it is deemed an organic part of its mother rather than an independent entity (Jakobovits, 1967). Furthermore, underlying Jewish principles assert that abortion is not murder, although it may amount to killing, with the difference lying in the circumstances. Killing is allowed in self-defense, when the victim is not innocent but is considered an "aggressor". A fetus may be regarded as an aggressor when the mother's health is endangered. Hence, the woman's interests, when health considerations are at stake, override those of the fetus (Steinberg, 1991).

The Jewish fetus has no "right to life" but a right to be born, but this right is relative to the welfare of the mother. In any moral clash, the mother has priority over the fetus, which is only a potential life (Feldman, 1998). Thus although Judaism accords a supreme value to life (Jakobovits, 1967), it permits abortion under

[19] This is true only since the mid-19th century (Luker, 1985)

certain circumstances and while some rabbis would justify abortions only when the woman's life is in danger, others are willing to extend religious exemptions when continuation of the pregnancy threatens the woman's physical or psychological wellbeing (Morag-Levine, 1994), as well as the future wellbeing of the child. In recent years, Jewish rabbis have issued different rulings regarding the question of abortion in a host of medical conditions such as Tay Sachs, Down's syndrome, anencephaly, X-ray exposure during pregnancy, maternal rubella during pregnancy and the like. For each of these conditions some have allowed the termination of pregnancy, while others have not (Steinberg, 1991). Barilan (forthcoming b), explains that according to Jewish law, a child's claim to life begins only upon the 30th day after birth and thus the threshold of viability is post-partum. In terms of *Halacha*, in order to claim full rights for the protection of life, one must have an established claim to the capacity to maintain independent and lasting life. Thus, Jewish doctrine does not stress biological viability alone, but viability in the sense of the baby acquiring viable attributes of *imago dei* (Barilan, forthcoming b). This delay in the attribution of full protection of life is true even post-natally, as regards premature neonates. Here, the value of protecting them from future suffering may in the first 30 days prevail over the value of life. Moreover, in Jewish law, the right to destroy a human seed before birth is entirely unrelated to the question of the entry of the soul before birth or to the claim to salvation after death, in contrast to other Western religions (Jakobovits, 1967).

The general framework of abortion policy in Israel subscribes to these principles. According to Amir (forthcoming), Israeli abortion law is actually Jewish law, in particular because it has two unique characteristics. First the law makes no differentiation between the mother and the fetus, which means that there is no time limit for any abortion. This is highly unusual compared to most other Western countries and particularly significant in the case of "late" abortions. The second feature is a clause that permits abortion on the grounds of extramarital sex, which is based on the Halakhic fear of giving birth to what religious law calls a "*mamzer*" (bastard).

Most of the Israeli counselors I interviewed declared themselves to be "actively" secular, meaning not only do they not obey religious rules but that they perceive them as contradictory to their scientific ways of thinking. However, their interpretation of mothers and fetus' rights was revealed to be not as far removed from the traditional Jewish view as they themselves believe.[20] As we have seen, in the interviews, Israeli counselors disregarded the fetus' rights and saw the fetus primarily as a part of its mother and not as an autonomous being. However, my claim is not that concrete religious practices and beliefs shape the counselor's views regarding

[20] What is surprising is that traditional attitudes apparently do not affect only Israeli (Jewish) counselors, but also American Jewish women. In Rapp's study on New York women who undergo amniocenteses, Jewish (non-orthodox) women were the most prone to abort after a "positive diagnosis", and to have what Rapp describes as a "user-friendly" attitude toward medical intervention (Rapp, 1999)

mothers, fetuses, "life" and abortions. Rather, religious ideas function as moral-mental structures, which linger on in contemporary society (Latour, 1993; Rabinow, 1999; Shenhav, 2005). As such underlying structures of belief, they thus frame the way the counselors interpret the moral dilemmas associated with reproductive genetics.

In sum, the Israeli stance in the abortion debate is deeply influenced by the Jewish interpretation of abortion, which is far more flexible than the Catholic one. In stark opposition to the German-Catholic view, the Israeli-Jewish fetus is not considered to be a "life" right from conception, and it has no rights. This difference is further strengthened by the fact that the Catholic Church has made abortions an important issue on its political agenda, while Jewish rabbis have not.

RELATED REPRODUCTIVE LAWS

All state laws and court decisions correspond to their legal environment. The Israeli legal context works to support abortions in general and selective abortions in particular, while the German legal context restricts the practice of abortion. Currently, the Israeli abortion law offers no rationale for the general prohibition it imposes on abortion, as there is no reference to the status of the fetus, its rights or those of pregnant women. Furthermore, there are no statutory or administrative parameters for the severity of the risk to the fetus or pregnant women that would justify an abortion, and no definition of the nature of the mental harm the law envisions (Morag-Levine, 1994). This lack of caveats concerning risk assessment, results in what Fink and Glick (1993) describe as overestimation of risk and a rush towards abortions. Furthermore, in Israel the abortion law is not associated with other laws that deal with fetal rights, as the Israeli fetus simply has none. However, the most important Israeli legal verdict in this area is one by the Supreme Court concerning a wrongful life suit.[21] Wrongful life suits permit the infant to sue medical agents and accuse them of negligently failing to detect a fetal anomaly, or informing the parents of its potentiality. As a consequence, the child (as well as the parents) can demand compensation, on the grounds of his life being "wrongful". Thus, the fetus' right not to be born has been acknowledged by the Israeli Supreme Court, which stated that in certain cases – those of "wrongful life" – non-existence is preferable to existence with disability (Jellinek, 1997). To put it bluntly, the only right the Israeli fetus holds is the right not to be born handicapped or the right to be born "healthy". This contrasts with Germany, where the courts repeatedly argue that life, even with disabilities, cannot constitute damage, since human life is a legally protected object of the highest value, and there is no "right of nonexistence" (Harrer, 1994). However, this does not mean German practitioners are not exposed to legal threats, as the German law accepts "wrongful birth" suits that entitle the parents to full compensation for the total burden of raising a disabled child whose

---

[21] Israel's Supreme Court: Zaitsov v. Katz, C.A.540/82, 40 P.D. (2) 85.

disability was not diagnosed due to negligence. Yet "wrongful birth" suits pose no legal, philosophical or moral problems, as they are classic claims in torts involving plaintiffs (the parents) who had the status of persons and thus held legal rights when the "wrong" was done. Additionally, they do not declare any life to be wrongful, or favor non-existence over existence with disabilities (for an elaborated discussion of the wrongful life issue, see Chapter 7). An additional difference between the Israeli and German legal environments can be seen in the most important German law dealing with the rights of pre-embryos formed at the laboratory outside the woman's body,[22] namely, the "Embryo Protection Law" (EPL),[23] which came into effect in January 1991. This law confers rights to very early fetuses, thereby limiting the use of new reproductive technologies in Germany. As a result of this stance, it prohibits the following: egg and embryo donation, attempting to fertilize more eggs from a woman than can be transferred to her within one IVF treatment cycle (three eggs according to the law), in order to prevent intentional embryo wastage, surrogate motherhood, preimplantation diagnosis, sex selection (in the absence of sex-linked inheritable diseases), fertilization with sperm retrieved post-mortem and cloning. Most of these techniques (except cloning) are not prohibited in Israel. The EPL is also the cause for prohibition of stem cell research in Germany, a scientific and moral issue that reached the headlines in Germany, with rival parties lining up in two camps according to their views on stem cell research as either extremely important for the future of German science or as a dangerous moral transgression.[24] My interview materials demonstrate to what extent the stem cell debate has shaped the counselors' answers to questions about abortions. While the interviewed counselors, scientists who often support stem cell research, were not directly asked about protection of very early embryos, many of the German interviewees still used this discussion as a cultural framework in which they articulated their opinions concerning the status of the fetus. This is very different from the Israeli context, in which protection of the early stages of pregnancy, namely stem cells or embryos, which result from IVF treatments and are outside the woman's body, is basically a non-issue, professionally, ethically and politically (Prainsack, 2006a).

In summary, the legal environment of the German abortion law grants rights to very early fetuses and does not accept "wrongful life" suits. Furthermore, in Germany there are inherent contradictions within the cluster of reproductive laws themselves (Richardt, 2003). In contrast, through its acceptance of "wrongful life" suits, the Israeli legal context supports selective abortions and pushes professionals to elect termination of pregnancy rather than taking any risks for the future health of the unborn child. Thus, whereas the German legal discourse concerning the fetus centers on concepts such as "life", the Israeli legal discourse centers on the potential risk of "wrongful life".

---

[22] Pre-embryos are also called "fertilized eggs", "preimplantation embryos", or "frozen embryos". The terms themselves hint at the controversy surrounding the moral standing of this relatively new, scientifically created entity.

[23] (1990) BGBI. I. S. 2747

[24] The stem cell debate was partly resolved in 2002 (see Lanzerath, 2004).

CONCLUSION

In this chapter I have demonstrated how personal bodies of mothers and fetuses are treated in light of the body politic of their nation or more specifically in light of ideas about the fetus' and mother's rights, the concept of "life" and society's morality and interests, which translate into the abortion policies and practices of both "normal" or "abnormal" fetuses, late or early during the pregnancy. In the case of Israel, it is quite easy to abort at all different stages of pregnancy, for diverse reasons but especially in the case of embryopathic indications. Moreover, in cases of uncertainty, in which the pregnancy may be at any medical risk, abortions are the default. In Germany women also abort in large numbers during the early stages of pregnancy and it is obvious that many selective abortions take place before viability, despite a lack of registration. Yet, a comparative analysis proves that German professionals attempt to restrict this tendency, especially when it comes to "late abortions". Thus, "late" selective abortions are far more common in Israel than Germany, despite the fact that since 1995 both legal systems permit them (although for different reasons). In this chapter I have shown that in order to better grasp these rates of "late" abortions, the ways in which professional gatekeepers such as genetic counselors and medical doctors implement the law and facilitate or hinder what it allows needs examination, as it might be misleading to only study the law itself. Furthermore, I have argued that in order to understand professionals' discretion, one must study the framing of the abortion debate within their respective societies.

Whereas in Germany the framing of the abortion debate centers around issues such as the sacredness of human life and what the treatment of embryos says about the morality of the nation, in Israel the fundamental questions are how risky the pregnancy is, does it involve the future suffering of the mother or child to be and how do abortions effect the demographic "threat", as protection of the embryo is a non-issue (Gross, 1999). In that regard it should be noticed that much of the framing and its practical impact is unconscious (Schoen and Rein, 1994). Hence, when a German interviewee says, for example, that she feels uncomfortable being involved in "feticide", it is very unlikely that she will explain her position to result from the fact she grew up in Germany, a nation with a collective trauma emanating from the fatal engagement of medicine and the Nazi regime. Similarly, Israeli interviewees might follow Jewish reason in their abortion practices – while at the same time declaring themselves to be totally secular and unaware of Jewish law. Nevertheless, I argue that framing matters. Thus, in the case of Israel, the flexibility of abortion, as reflected in the professionals' attitudes and practices documented in this chapter, can be best explained as a result of a confluence of factors that frame the abortion debate and include Judaism, the Israeli abortion law's non-problematic history and the law's non-restrictive political-legal context. Moreover, in Israel the mother or parents are considered to be the professionals' clients and the fetus is not believed to have any rights, since it is not considered a "life". The contrary tendency of German medical gatekeepers to hinder "late" abortions can best be explained by a combination of factors, including the effect of Catholic doctrine, the controversial history of the German abortion law, a special sensitivity towards selective abortions (due to their historical connotations) and a particular understanding of the fetus as a "life" and thus as a symbol for the new moral Germany (Wuerth, 1997).

The similarities and differences between the Israeli and German abortion laws take on a different light from a feminist perspective. In both countries the power of the state and doctors over women's lives is very strong and abortion is not a right but a favor for those in need. In other words, abortion is not accepted in either of the two countries, it is merely tolerated. Both abortion laws are restrictive in nature, as abortion remains a criminal offence unless it is authorized under specific conditions. The two abortion laws are also built around a similar basic contradiction: they view women as designed by nature to be mothers and therefore ethical, life-preserving beings. On the other hand, the fact that women abort seems to prove their incapacity for ethical judgment (Amir, forthcoming). This idea is turned into practice by the obligatory counseling that exists in the two States. As stressed by Funk (1993), the hidden assumption behind obligatory counseling is that women are not morally competent to be trusted to decide for themselves whether to have an abortion or not. In this sense, both German and Israeli abortion laws disregard feminist motivations for abortion and the basic right of a woman over her body. In Germany feminists have been fighting over this issue quite unsuccessfully since the 1920s, whereas Israeli feminists have basically abandoned this troubling issue altogether. Yet concerning abortions on embryopathic grounds, German and Israeli women who discover medical problems in their fetus late in pregnancy and wish to abort, face a different ordeal. This is true despite the fact that abortions due to the fetus' medical condition are allowed in practice until the very end of pregnancy in both countries. The Israeli woman is likely to get sympathy from the medical system and her request is also explicitly acknowledged by the law as a just cause for abortion. The German woman needs to place the blame on herself and claim she is mentally unable to raise a disabled child. The German woman is also likely to run up against several refusals before she finds a doctor or hospital that agrees to perform the termination or she might even have to travel abroad in some cases.

CHAPTER 6

SEX CHROMOSOME ANOMALIES (SCAs) IN ISRAEL AND GERMANY: ASSESSING "BIRTH DEFECTS" AND MEDICAL RISKS ACCORDING TO THE IMPORTANCE OF FERTILITY*

WHAT ARE SCAs?

As we have seen, Israeli and German counselors differ in their positions towards the definition of a whole range of medical conditions in the fetus and as regards the optimal practical outcomes of these definitions. This chapter focuses on a group of such conditions, namely sex chromosome anomalies (SCAs), to highlight how cultural beliefs concerning the "human natural purpose" (Canguilhem, 1991), or more specifically reproduction as a human norm, can lead to differences in the counselors' differential risk assessments and definitions of the "normal" and the "pathological" that form the basis of their practices. SCAs are a collection of atypical diagnoses involving too many or too few sex chromosomes. The specific (and most common) SCAs addressed in this study are Klinefelter syndrome (XXY) and Turner syndrome (X0 or 45X). Males with Klinefelter syndrome have an extra X chromosome. They are tall and infertile. In rare instances they may also be mentally retarded, though the majority of people with Klinefelter syndrome lead completely normal lives and their condition is only discovered when they wish to have children (Vardimon and Ben Rafael, 1999). Women with Turner syndrome have only one X chromosome (instead of the usual two) and thus 45 instead of 46 chromosomes. Many women with Turner syndrome are relatively short. All are infertile. In some cases, the syndrome involves other physical irregularities, such as heart problems. In girls the condition is typically diagnosed at puberty, when it has become clear that they have not started to menstruate (Vardimon and Ben-Rafael, 1999).

Today, sex chromosome anomalies are part of a range of potential findings in CVS and amniocentesis that have become more frequently identified as an unintended consequence of testing for more severe conditions (mainly Down's syndrome). Finding such anomalies is a typical case of technological inertia or the "snowball effect" in which one medical procedure leads to another and thus to the emergence of unexpected problems and dilemmas. Hence, in such cases knowledge can be a burden, since without these tests children with sex chromosome anomalies would be likely to go through a large part of their lives with their "problem"

 A version of this paper has been previously published in Medical Anthropology Quarterly.

unnoticed by parents and doctors as well as themselves. However, after a "positive diagnosis" during pregnancy of a problem that would unfold in adolescence, it is no longer possible to go back to a state of innocence, since the institutionalization of knowledge and the routinization of tests have by then created an irreversible situation. Furthermore, SCAs are neither preventable nor treatable. The fact that, by and large, they do not affect intelligence, life expectancy, physical appearance or even sexual preference, makes them a distinctive borderline case and sets up a difficult moral dilemma. In the following, I will argue that my findings concerning the ways in which the professional groups under study think about risk and about SCAs show them to be embedded in their local morality (Kleinman, 1995) and definitions of the natural human purpose, and not in their medical expertise or the ethics of their profession, which centers on the concept of non-directivity.

STATISTICAL PROPS AND MEDICAL RISKS

A central feature of late modernity is that lay actors as well as experts tend to think in terms of risk assessment, as part of a general climate of risk (Beck, 1992; Giddens, 1999). In medical genetics, a scientific style of reasoning has been based on probabilities ever since its early days, when Sir Francis Galton (Charles Darwin's cousin) founded both the biometrical school of statistical research and the eugenic movement. Hacking (1990) has argued that the phenomenal rise of statistical reasoning can lead to an over reliance on probability in the determination of values. Deviating from Hacking, I argue that despite the fact that genetic risks are commonly statistically assessed, in the present case statistical probability was not found to dictate values and nor did computations replace ethical judgments. Rather, they functioned as a rationalized justification for *a priori* moral positions, because in genetic assessment, the numbers and computations themselves are meaningless without value judgments about questions such as what constitutes a "high" or "low" risk, or a "worthy" or "unworthy" life. Statistical estimates, whatever the field, are never clear cut and their interpretation can change across cultures and time. For example, Katz-Rothman reports that in the field of medical genetics in the US, risk rates which during the 1960s were considered low, came to be perceived as high during the 1980s (Katz-Rothman, 1986). What is it then, that makes the very same numerical risks seem "high" at one point in time and "low" at another, or severe enough to justify abortion in one society, while not so in another? Sociological and anthropological wisdom has taught us that risk can best be understood as a social construct. The cultural/symbolic approach to risk (Lupton, 1999) put forward by Mary Douglas, who wrote a series of influential books on the subject (Douglas, 1966, 1970, 1986, 1992; Douglas and Wildavsky, 1982), argues that risk itself is a socio-cultural concept, and that perceptions of risk are communal. Therefore, according to Douglas, medical risk assessments should be seen as cultural reactions to sins, moral transgressions or the breaking of social taboos and not as "objective" medical interpretations. Douglas and Wildavsky (1982) note that public fears about risks are ways of maintaining social solidarity and do not reflect "real" health or environmental concerns. Thus, risk judgments are political, moral and aesthetic because they are connected with legitimating principles or logics of justification (Boltanski and Thévenot, 1999, 2006; Resnik and Frenkel, 2000) that affect the ways they are perceived. In the following, I will demonstrate how German and Israeli counselors' interpretations of "calculated risks" (in relation to SCAs) are mediated by their respective cultures' understandings of the "normal" and "pathological" or of "the human natural purpose", which Canguilhem (1991) sees as fundamental to any definition of normality. Thus, the following data will demonstrate that although there is a biological reality to risk assessments, they take on specific forms in different cultures.

The range of syndromes that can accompany SCAs is wide, including risks of learning disabilities, mild mental retardation, growth and stature anomalies and, in the case of Turner syndrome, atypical neck and finger formation, and possibly heart problems (Vardimon and Ben-Rafael, 1999). Yet, in the interviews it was surprising to find that there was no agreement among counselors in either nation about what SCAs "really mean", or the risks they entail. While some mentioned a statistical reduction of 5–10 points in I.Q. (in the cases of triple X or Klinefelter syndrome) or a tendency towards aggression (in the case of XYY), others dismissed such talk, saying it was based on scientifically spurious and old-fashioned theories. Nor was treatment widely agreed upon: while some experts said that today's fertility treatments can offer a solution for the infertility involved in some SCAs (Klinefelter syndrome, for example), others saw this as possible in rare cases alone or in the distant future. Psychological assessments of the difficulties of living with such conditions also varied significantly. For the most part, Israeli counselors tended to emphasize the accompanying risks of SCAs, while German counselors played them down.

COUNSELING FOR SCAs IN ISRAEL AND GERMANY

The views of genetic counselors in Israel and Germany regarding aborting for Turner and Klinefelter syndromes are reported in Table 1 (aggregate data from the two surveys).

Counselors were also asked how they personally would cope with the discovery of Klinefelter or Turner syndromes in the first trimester of pregnancy. Their answers are presented in Table 2.

Tables 1 and 2 show very clearly that Israeli counselors are far more supportive of selective abortion in the cases of Klinefelter and Turner syndromes than their German counterparts, in both their professional and personal behavior.[1]

In order to learn more about the reality of their work, during the interviews I asked the counselors to describe problematic (and non-problematic) counseling sessions,

[1] No significant differences were found between attitudes towards Klinefelter versus Turner syndromes in either Germany or Israel.

TABLE 1. How Would You Counsel in the Cases of XXY and 45,X?

	XXY		45,X	
	Israel	Germany	Israel	Germany
Urge to carry to term	–	15%	2%	18.3%
Emphasize positive aspects	13.7%	42.9%	4%	43.2%
Be as unbiased as possible	68.6%	41.3%	74%	37.3%
Emphasize negative aspects	17.6%	0.4%	18%	0.8%
Urge termination	–	0.4%	2%	0.4%
	100%	100%	100%	100%
N	53	242	53	242

TABLE 2. How Would You Personally Act in the Cases of XXY and 45,X?

	XXY		45,X	
	Israel	Germany	Israel	Germany
I would abort	84%	13.6%	88%	14.3%
I would not abort but it should be legal	14%	66.7%	10%	69.6%
I would not abort and it should be illegal	2%	19.7%	2%	16.1%
	100%	100%	100%	100%
N	53	242	53	242

in which abstract moral values confronted mundane professional decision making or parents' fears and distress. It was this question that often led counselors to talk about SCAs, thus directing my attention to it. However, whereas German counselors tended to offer SCAs as an example of a finding that must be revealed to patients (for legal reasons) but that does not justify abortion because it *only* affects fertility, in Israel SCAs were cited as an example of a troubling case because of the lack of sufficient professional clarity regarding possible accompanying risks. The German counselors' discomfort stemmed from the fact that their patients' first reaction was fear at the discovery of a genetic anomaly. Most of the German counselors viewed their role here as comforting the future parents and explaining to them that their child would be "fine", albeit infertile; the German counselors never interpreted infertility in itself in terms of a pathology or disability. For example, a (male) German genetic counselor and gynecologist who heads a large private clinic, said:

"An especially difficult case for counseling is sex chromosome anomalies. For example, when we discover XXY Karyotype, we know it will be a non-fertile boy but without mental retardation. But many times the parents demand an abortion in this case, which I don't see as justified. I refuse to make a recommendation and it is a conflict situation. Some parents accept my position that since there is no mental retardation, abortion should not take place but not all of them do so. In many cases I give my recommendation according to what the mother says she can or cannot stand, but not with Klinefelter syndrome".

Concerning fertility matters in Germany, the same counselor said:

"Reproductive medicine is not a highly developed field in Germany because we don't care so much about reproduction. There is no fertility craze here and this whole medical field is unfashionable. Anyway, 1/3 of German women choose not to have children. I think the explanation is that we lack a public sense of future. We have a lot of past, some present, but no future. Families are not supported and people are afraid of losing their job if they choose to become parents. Everything here is too relaxed, we climbed so high after WWII that now we are just a hedonistic society, a fun society, with no future and no children".

A woman MD counselor working in a hospital said she did not perceive the finding of future infertility itself to be problematic; rather what troubled her was the fact that the parents were not properly prepared to deal with such information.

"A difficult case for me is when there is an unexpected diagnosis, when the parents did not get the information beforehand or got the wrong information. For example, sex chromosome anomalies that were discovered in amniocentesis without prior counseling. To the parents it sounds frightening but usually when I explain to them what it really means, they don't see it as a reason for abortion. I personally think that when no other physical or mental problems are involved, infertility is no reason for abortion. Still, I accept that some cases are exceptional and if the parents really don't want this child, I do not see it as a major problem and accept their wishes. In twenty years of working experience this has happened to me just twice, once for 45,X, and once for Triple X".

Genetic counselors in Germany consider it to be a professional and moral success if they can prevent abortions for SCAs. For example, a woman counselor and gynecologist working in the private sector said:

"With sex chromosome anomalies, I try to be as unbiased as possible. I provide addresses of parents with such children and of support groups. And luckily, in my case, no one has stopped a pregnancy because of Klinefelter syndrome and I have had about six such cases in my career. Turner is another story because it can involve other problems in addition to the infertility. But I know many women with Turner syndrome and they are very happy. Sometimes they suffer from social stigma because they look a little different but that's all. And in any case, I surely wouldn't define infertility as a disease".

Another woman, a general practitioner and private counselor, used a similar argument that infertility should not be interpreted as pathology and cited the quality of life of women with Turner syndrome:

"Infertility is not a disease and surely not a disease one should give up living for. Anyway, so many people don't have children... Turner syndrome is a little more difficult because it might involve other medical problems. And yet I know many happy women with Turner syndrome and I am sure they would not wish to be unborn."

The German counselors' logics of justification regarding selective abortions are also institutionally grounded. German genetic counselors can make recommendations for selective abortions but the final decision lies with the hospitals and doctors who perform the procedure. Thus, the counselors' knowledge of hospital policy towards different reasons for abortions is another factor behind their reluctance to support selective abortions due to SCAs. As described by a pediatrician and counselor working in a city hospital:

"When we see XO or XYY Karyotypes, I tell the parents all about this problem. For example that hormone therapy is possible. But I tell them that the infertility cannot be cured. A few parents seem

frightened but the majority wants to continue the pregnancy. And in any case, it is hard to abort for these reasons in Germany. For minor problems like this or like a cleft lip or a treatable heart problem, it is hard to find a hospital that would be willing to perform the procedure."

Some German counselors even perceived the "burden of knowledge" as a positive contribution of their practice to these children's future wellbeing. For example, the head of a genetic unit in a large hospital said:

"Many people are born infertile. But if they are informed from early childhood, I think it can even be an advantage. Like anything from the tree of knowledge, it has advantages and disadvantages. With this knowledge, they can plan their lives differently. Anyhow, I don't see infertility, which might even have a solution in twenty years, as a medical problem, since it does not affect mental development".

Others were aware and subtly critical of foreigners coming from a different culture who do not tolerate SCAs. For example, a counselor working in the private sector said:

"I would not dream of aborting for sex chromosome anomalies but I have had patients who traveled to Holland in order to abort a girl with Triple X. They came from Turkey and I tried very hard to convince them that this is no problem but they did not want the child".

German genetic counselors thus view Klinefelter and Turner syndromes as examples of a non-pathological difference and they therefore play down the risks of accompanying problems. In comparison, when Israeli counselors discussed SCAs they did not stress the infertility connected to such anomalies but instead emphasized the links to more severe conditions. The additional medical risk (over and above infertility) involved in such conditions, which was minimized by German counselors, especially when discussing Klinefelter syndrome, was emphasized by their Israeli counterparts, who complained about the lack of clear professional knowledge about the risks involved in such conditions. For example, a Ph.D genetic counselor working in one of Israel's major hospitals said:

"Sex chromosome anomalies are in the grey area. XYY, XXY or X0 might be completely normal or might have lower intelligence. Counseling in such cases is not hard for me ethically but rather because the couple may find it hard to reach a decision and because the numbers are not clear. It is not black and white. But I completely understand it when people choose to abort for these reasons. I don't think it's good to raise a child under a magnifying glass and worry all the time about what will become of him or her. I myself would abort in those cases, so why should I stop others? I think that if it is an aberration from the norm, it's a problem, even if there are almost no clinical manifestations".

In a similar fashion, the head of a hospital genetic institute, a pediatrician and a counselor, raised the issue of the (paltry) state of medical knowledge concerning SCAs.

"I feel counseling is easy when the medical literature tells me exactly what my finding means or when the woman knows ahead of the finding itself how she would act in such a case. Difficult situations arise when I don't have enough information about what the finding really means, like in the cases of XXY, XXX or XO. But I don't really see why people should take the risk of a mental problem or learning disabilities when it can be avoided. And on top of all this, infertility may not be strictly defined as a medical problem, yet socially and psychologically, it surely is."

Even Israeli counselors who felt uncomfortable about aborting for SCAs, or about deciding not to take the "minor" risks associated with them, did not see it as part of their role to talk the patients out of their decisions to abort. A hospital-based genetic counselor (a woman) explained this as resulting from her non-directive counseling:

"My patients are usually interested in every possible technology, without critically considering where it might lead them. For example, only very few couples who go through amniocentesis continue pregnancies with even very minor problems. Most pregnancies with Klinefelter syndrome, for example, are interrupted. People don't want to find out what it really means. They just don't want to take any chances. They want to make sure they have a "healthy" baby. But even if I dislike their attitude, it is still their baby and I am not going to argue with them whether infertility is a major or a minor problem, physically or emotionally or whether a 5% risk of learning disabilities is a major or a minor risk".

Another woman counselor working in hospital, who felt uncomfortable about selective abortions for SCAs, justified them by referring to her perception of intolerance in Israeli society:

"I find it hard when people abort for Klinefelter or Turner syndromes, since they can be very fine people and without our tests those abortions would never take place. The very small risk of mental retardation or learning disabilities is in my opinion no reason to abort. But we have to think of the society these children will grow up in. Israelis are intolerant as regards genetic problems or infertility and that's why I accept the parents' wishes. I can't expect them to swim against the social tide on their own".

This counselor's logic of justification illustrates the vicious circle where Israeli counselors perceive their society's intolerance towards the infertile and the disabled as an independent variable, which then accounts for their conduct. They fail to perceive intolerance as a cultural trait that they have a hand in shaping as experts in this delicate field of practice and knowledge.

Surprised that infertility itself was not often discussed in the interviews in Israel but rather other accompanying risks, I asked a direct question about the infertility caused by SCAs during a lecture I gave to Israeli counselors in their professional rounds. Before starting my talk, I asked the audience to write down how they personally would react to a pregnancy with Turner or Klinefelter syndrome and how they interpret infertility. All the respondents said they would choose to abort because infertility was a disease or "medical handicap". Although the accompanying risks were mentioned, the majority of the participants at the meeting thought that infertility on its own was a strong enough reason to abort. Typical answers were:

"There are enough problems in life anyway. Therefore I would not knowingly have a child with a physical, social and psychological handicap".

Another counselor based his attitude on a differentiation between pre- and post-natal disability:

"I differentiate between the definitions of handicap before and after birth. I would not say infertility among adults is a disease but in the fetus, combined with higher medical risks for other problems, I do see it as a reason for abortion".

Another counselor simply wrote:

"Infertility is a disability. And when you add to it aesthetic problems and risks of cognitive problems, I would surely abort".

Only one of the participants, who had been experiencing difficulties getting pregnant and was at the time undergoing IVF treatments, said she was not sure how she would act if she was pregnant with a fetus with XXY or X0.

Thus in the case of SCAs, risk assessments emerge here as the outcome of cultural embeddedness and the ways in which communal concerns and values form the link between private and collective bodies. As the data show, in contrast to their German counterparts, Israeli counselors perceive infertility as a major medical and social problem. Equally, Israeli counselors also stress the potential additional risks of SCAs to a far greater extent than their German colleagues. In order to understand how it is that professionals with the same medical knowledge in two post industrialized societies interpret the same scientific facts so differently, the cultural significance of fertility in each of the two nations needs to be examined.

THE IMPORTANCE OF FERTILITY IN ISRAEL AND GERMANY

Alex Von Cube (1989), a German researcher who studied the fertility decline in Germany wrote in a biographical note: "From adolescence through adulthood the debated questions included whether one should or would have children one day or at all and if so, under what circumstances those children would live" (p. 5).

On a similar biographical note, I would add that despite the available (medical) option in Israeli society to choose not to have children, despite it being an intriguing philosophical question, which may be typical of adolescent challenges to social norms and even with the fact that it is unclear under what circumstances future Israeli children will live, I personally have never heard this question raised among my friends. On the contrary, as Israelis, we thought of parenting as a crucial and indisputable expectation we have from our lives and ourselves.

In this section I situate genetic counseling for SCA in Germany and Israel within a broader cultural context, the fertility culture of both societies, in order to explore the hypothesis of a deeper socio-cultural explanation for the findings. I review the motivations for becoming parents in Germany and Israel and their relation to governmental policy, history, religion, the status of women and economic and demographic fears. The total fertility rates (born/woman) in the two countries is a revealing place to begin. In Israel the estimated birth rate at the time of this research (2001) was 2.57, whereas in Germany it stood at 1.38, below replacement level (C.I.A World Fact Book, 2001). Various writers have shown Israel to be a very family-oriented society, with low divorce rates and high rates of marriage and birth compared to other post-industrial countries (Berkovitch, 1999; Fogiel-Bijaoui, 1999; Peres and Katz, 1991; Portugese, 1998; Remennick, 2000; Ivry, 2004). On the other hand, although the fertility decline in Germany is part of a general tendency in industrialized nations, the steepness and the duration of the

German decline have distinguished it even from other nations that have converged at a level below replacement fertility (Von Cube, 1989). In West Germany, sub-replacement fertility levels were reached much earlier than in other countries with low fertility rates and the FRG is often cited as a prototype of fertility decline in the industrialized world. This tendency reached its peak after unification, when, as noted by Eberstadt (1994), Eastern Germany's adults have come the closest to a temporary suspension of childbearing, compared to any other population in human experience. Such low fertility rates can be accounted for by small family size but also by the large number of people who chose to remain childless. According to Dorbritz and Schwarzer (1996), childlessness became the most crucial factor behind Germany's demographic trends and forecasts suggest that 30% of women from the former West Germany will never become mothers.

These contemporary fertility tendencies in both societies fall into sharper perspective if we examine Israel and Germany's fertility policies in the 1950s, the founding years of both Israel and the two Germanys in the post-WWII era. In the fifties, as today, these countries were worried about the size of their populations and thus maintained pro-natalist policies (Berkovitch, 1999; Moeller, 1993). However, in Germany this was mostly to do with the needs of the economy and the social security system, while in Israel it was (and still is) mostly due to a perceived ethnic threat (Kahn, 2000; Kanaaneh, 2002; Weiss, 2002). In Israel the "demographic threat", i.e., the fear that the Jews will not make up the majority of Israel's citizens, defined the political agenda of the Zionist society right from the start and led to the encouragement of fertility and the nationalization of the maternal womb. The declaration by Ben-Gurion, Israel's first prime-minister, that 100 Israeli Liras (currency at the time) would be granted to "Heroine Mothers", who had ten children and more, is an eminent example of this tendency (Berkovitch, 1999). West Germany was also troubled by demographic threats in the 1950s but ones that emanated from a different source. Adenauer, the German Chancellor at the time, explained this concern in his opening remarks to parliament, stating that West Germany was a rapidly aging nation, a fact which threatened the future of economic growth and stability of the social security system. Adenauer was worried that more people were leaving the workforce than entering it, as he believed only a constant birth rate and not machines were the best guarantee of social prosperity and security. Therefore, he attempted to strengthen the family and hence, the desire for children (Moeller, 1993). In East Germany, immigration to the West threatened the size of the population and its ideological strength and was among the reasons for the construction of the wall between the two Germanys. Thus, in all three countries in the 50s, political worries about the size of the population provided justifications for supporting parenthood, albeit for different reasons.

However, as the dissimilar birth rates suggest, pro-natalist governmental policy alone cannot explain citizens' willingness to become parents. Indeed, in Germany, statistics suggest that pro-natalist measures did not affect the long-term tendency towards smaller families (Moeller, 1993). By contrast, fertility rates in Israel have remained fairly high, although researchers disagree about the nature of Israel's pro-natalism (Melamed, 2002; Portugese, 1998; Schiff, 1981).

Importantly, these debates show that government policy cannot sufficiently explain different societies' birth rates, in that pro-natalist policy is not successful without a supportive cultural background, such as found in Israel. Hence, I argue that in order to clarify this issue, one must conduct an analysis which combines the political and social-cultural spheres but without allowing either one of them to displace the other. I shall now offer such an analysis.

When trying to understand Israel's high fertility rate, it can best be understood as deriving from a complex combination of causes – over and above government policy. These include identification with the collective goal of fighting the "demographic threat" and Arab enemies (Yuval-Davis, 1987), the threat of losing a child in war or in a terrorist attack and Jewish genocide during the Holocaust, which led to a wish to replace the dead by having many children. A further factor is Jewish tradition, in which parenthood is an important moral commandment and infertility is treated as a severe disability (Marx, 2002). According to Kahn (2000), the theme of women's infertility is the archetype of suffering in the Israeli/Jewish imagination. The Bible is replete with stories of barren women, like Sara or Hannah because in Judaism, a woman who cannot become a mother epitomizes the essence of female suffering. Kahn (2000) also points out that Jewish women are the primary agents through whom the nation can be reproduced as Jewish (this stands in opposition to missionary religions, which can recruit believers from outside the group) and hence Israeli fertility culture also stresses the importance of a genetic offspring. Childlessness is equally tragic for the Jewish man, who is commanded by religious law to reproduce ("be fruitful and multiply"), a duty beholden upon all members of society, including its highest religious authorities. Indeed, the ideal of celibacy is strikingly absent from Judaism (Jakobovits, 1967).

What can explain the opposite trend in Germany? Departing from largely economic and modernization theories, which are not specific to Germany, demographic theories have examined changes in values regarding children and the motivations behind fertility decision-making in Germany. Choosing to remain childless, or "childfree" (a term which has no equivalent in Hebrew), is normative in contemporary Germany, as it opens up opportunities for higher material consumption and ease of travel for couples and individuals, who are not stigmatized for their choice. The fact that childlessness is a normative option in Germany was also clear from my interviews with German counselors. For example, many suggested that couples who knew that they carry a genetic disorder but would not like to abort selectively, could choose not to become parents at all. In contrast, Israelis never mentioned giving up on parenthood as a possible solution to such cases.

On the other hand, alongside the fact that childlessness is normative in contemporary Germany, women who do decide to become mothers are faced with a social structure that makes mothering very difficult. This results from traditional ways of thinking about the role of women, which are still very prevalent, with the consequence that it is very difficult for a mother to sustain a career (Blossfeld and Rohwer, 1995; Gauthier, 1992; Spakes, 1995).

While the fact that women's identity in Germany, either as wives and mothers or as "childfree" career women, pushes German women away from parenthood, unlike in Israel, religious affiliation or belief do not pull in the opposite direction. In fact, religious affiliation or beliefs have been found to be unrelated to fertility among German women (Dorbritz and Schwarzer, 1996). According to the Family and Fertility Survey,[2] cited by Dorbritz and Schwarzer (1996), German childfree women rank their reasons for not having children as follows: 1. It is difficult for women with children to work 2. Raising children entails suffering and problems. 3. It is expensive to raise children. 4. Today's living standards are not suitable for large families 5. When you have children, there is no time left for other important things in life. 6. Pregnancy, giving birth and taking care of children are heavy burdens for women.

Looking for a German specific causation for low fertility, Von Cube (1989) argued that the German population has a fear of the future which emanates from two politicized issues: the threat of nuclear conflict and environmental degradation, which prompt German couples to have fewer children. However, it is difficult to grasp how an abstract fear of the future (in Germany) can negatively impact on fertility, whereas more concrete existential fears in Israeli society generate exactly the opposite response.

FERTILITY AND NEW REPRODUCTIVE TECHNOLOGIES (NRTs) IN ISRAEL AND GERMANY

The medical field of new reproductive technologies (NRTs) and the legislation accompanying it also shed light on attitudes towards fertility, since nations vary in the way their legal systems interpret the ethical dilemmas embedded in these technologies (Shalev, 1995). In Israel, reproductive laws and fertility services reflect the fact that parenthood is highly valued. According to Kahn, in the mid 1990s there were more fertility clinics per capita in Israel than in any other country in the world (4 times the number per capita in the US). Furthermore, Israel is extremely permissive in implementing NRTs and it enables any woman who wants to have her own biological children to do so with the assistance of medical institutions, free of charge, even if she is unmarried or lesbian, a policy which is a world exception.[3] Furthermore, the number of IVF treatment cycles until a second child is conceived is not limited by the Israeli health insurance basket, whereas in Germany

[2] The FFS (Family and Fertility Survey) was initiated in the late 1980s by an informal workgroup for population issues of the Economy Commission of the United Nations for Europe (UN/ECE). The German FFS was conducted in 1992 by the Federal Institute for Population Research.

[3] Carmel Shalev (1995) attributes this liberal stance to concerns on the part of Jewish rabbis that ovum donation may lead to the birth of *mamzerim* (bastards) according to religious law. Therefore ovum donation was initially authorized only for unmarried women who were undergoing IVF treatment.

only four cycles of treatment are covered by the national health insurance. Ruth Landau (1996), who analyzed the conclusions of the Committee to Examine *In Vitro* Fertilization in Israel, notes that there is a degree of asymmetry in the committee's conclusions regarding the right of a person to have a child versus the right of the child to parents; in fact, the Israeli committee placed the right to parenthood as superior to the right to have parents. Additionally, in 1996 Israeli legislators passed the Embryo Carrying Agreements Law, which made Israel the first country in the world to legalize surrogate motherhood (Kahn, 2000; Teman, 2003).

The German situation is strikingly different. The German "Embryo Protection Law" (1990), which came into effect in January 1991, does not support NRTs parenthood and actually works to limit possible reproduction. This law aims at limiting the use of NRTs in Germany. It highly protects the very early stages of pregnancy and confers rights to very early embryos. As a result of this stance, a number of technologies are prohibited, including egg and embryo donation, attempting to fertilize more eggs from a woman than can be transferred back into her within one IVF treatment cycle, in order to prevent intentional embryo wastage, surrogate motherhood, pre-implementation diagnosis (PGD), stem cell research, sex selection (in the absence of sex-linked inheritable diseases), fertilization with sperm retrieved post-mortem and cloning. Except for cloning, none of these techniques are totally prohibited in Israel and in fact they are warmly endorsed by the public. This, of course, allows for more children to be born as a result of the use of NRTs in Israel than in Germany. In Germany, however, there is no public opposition to the restrictions on NRTs, even though they limit citizens' chances of becoming biological parents. In fact, such opposition as does exist to the embryo protection law in Germany only relates to its restrictions on scientific research with stem cells and ignores its effects on fertility.

CONCLUSION

In this chapter I claimed that counselors' discourse in Israel and Germany on SCAs emerges from local understandings of biology and specific historically-shaped logics of justification (Boltanski and Thévenot, 1999, 2006). This is because the definition of the "normal" and the "pathological" in medical discourse is not simply "scientific" and "objective" but rather socially constructed. Thus, to affirm the "human natural purpose" by defining the abnormal, implies identifying the normative character of what is considered the "normal" state, and a support for certain values over others (Canguilhem, 1991). This chapter has shown that genetic anomalies are not "natural" in the sense of being above, beyond or deeper than culture. Hence, the use of genetic knowledge regarding SCAs in both countries varies dramatically, according to local understandings of reproduction as a human norm. Whether SCAs are perceived as a medical pathology or as a normative variance is related in both cultural spaces to assumptions about the importance of fertility in human life. Whereas in Israeli society parenthood and the reproduction of the species, or more accurately of the national womb, is understood to be what human life is all about, German adults do not share this assumption. Therefore, the practices of Israeli counselors reflect the demand not to deviate from the social and by extension the medical norm of fertility, whereas German counselors, who perceive the human natural purpose differently, do not see infertility as a social and thus a medical deviation.

The different moral understandings of this issue are also reflected in diverse local risk assessments. This is because the cultural and institutional filters through which risks are perceived constantly distort the probabilities (Douglas, 1986), in that genetic counselors' risk assessments are not strictly bio-medical but are rather strongly influenced by culturally-specific notions of the "normal" and the "pathological". Israeli counselors see future infertility in itself as a risk factor and stress the potential additional risks involved in SCAs (over and above infertility). In contrast, their German counterparts downplay both the additional risks and the importance of (in)fertility itself.

How does this difference affect counselees in Israel and Germany? Individuals who undergo genetic testing have been repeatedly reported to feel that their freedom to choose between different options is limited in a variety of ways. For example, Katz-Rothman (1986) observed that the choices available to women after amniocentesis are so constrained that the tested women experience themselves as being trapped by the knowledge they have been given and the responsibilities it entails. Nevertheless, the professional and social pressures put upon potential parents in Israel and Germany are very different. That is due to the fact that genetic counselors in Israel consider infertile life to be unworthy of living, in stark opposition to German counselors who adhere to a broader definition of a worthy life, one which includes the voluntary as well as non-voluntary infertile. There does not appear to be conflict between counselors and their counselees in either countries, as they seem to share the same cultural assumptions about fertility and (un)worthy lives.

CHAPTER 7

"WRONGFUL LIFE", IN THE EYES OF THE LAW, THE COUNSELORS AND THE DISABLED

New reproductive technologies have met with different legal responses in different countries, depending on their legal traditions and historical experiences, which affected their social perceptions of the meaning of "life", in all its different forms and stages. These new reproductive laws or rulings affect reproductive genetics in a host of ways. Chapter five has dealt with selective and "late"-selective abortions and with the interaction between the legal system and the practice of such abortions in Israel and Germany. The other major body of law affecting reproductive genetics is that dealing with "wrongful life" and "wrongful birth" suits. These legal claims are the focus of the first part of this chapter, which looks explicitly into the question of a life (un)worthy of living, as it is formulated by the two legal systems' handling of "wrongful life" and "wrongful birth" suits.

Turning to the genetic counselors in both studied groups, the chapter looks into their differential practices and examines them in the light of prevalent understandings of the question of "wrongful life" in their respective societies. As in the case of abortions, the legal situation concerning the issue of "wrongful life" in Germany and Israel is symbolically very different, but practically quite similar, as counselors in both states are exposed to very similar legal hazards. Yet, once again it is revealed that the society's symbolic regime strongly affects the counselors' beliefs and conduct, over and above the actual legal threat. After discussing the counselors, the chapter shifts its focus to the positions of the disabled themselves with regard to the notion of "wrongful life", and to their effect on the field of reproductive genetics, in both societies.

WHAT ARE "WRONGFUL LIFE" AND "WRONGFUL BIRTH" SUITS?

Prenatal genetic diagnosis, allows couples to avoid the birth of a fetus whose future life could be hampered by severe or minor physical and mental handicaps. This possibility has led to new kinds of medical malpractice suits named "wrongful birth" and "wrongful life". In order to discuss the court verdicts concerning "wrongful life" and "wrongful birth" suits in Israel and Germany, let me start by explaining these legal concepts.

"Wrongful Life" suits permit the infant to sue medical agents (mainly genetic counselors and gynecologists) and claim compensation for alleged negligence in their diagnosis, which did not sufficiently detect fetal abnormality or informed the future parents of its potentiality. Thus, the parents did not use the option of abortion and the child, who should have not been born, came to life. As a consequence, the child whose life is now declared to be wrongful, can receive compensation.

"Wrongful Birth": wrongful birth suits are different from wrongful life suits, as they are brought by the parents, against medical agents. The parents claim that they would have terminated a specific pregnancy, had they been properly informed of its risks or of a specific diagnosis, in an appropriate and timely fashion. Lacking this information, they were deprived of the right to terminate the pregnancy and avoid the child's birth.

What is the difference? Wrongful birth claims pose no legal-philosophical problems, as they are good old-fashioned claims in torts, involving plaintiffs (the parents) who had the status of persons and thus held legal rights, when the "wrong" was done. However, this is not the case with wrongful life claims, which pose great legal, conceptual and moral difficulties (Mason, 1998). As a matter of fact, such claims have been declined by most courts around the world (Heyd, 2002; Mason, 1998) for the following reasons (The following summary is partly based on the work of Harrer, 1994):

1. Causation: only in the most exceptional circumstances would the counselor have caused the defect and thus, there is a lack of injury. The counselor is responsible only for the consequences that followed the diagnosis, namely the non-interruption of pregnancy but his/her negligence cannot be said to have injured the child. Thus the counselor may argue that because of her action, even if negligent, the child is alive. This by itself holds a positive value and therefore, the counselor owes the child nothing. Moreover, nothing that the counselor might have done could have prevented the problem, since a cure was not available.
2. Comparison: the consequences of this, namely the born life, can only be compared to non-existence, which is the single available alternative, since "healthy" life was never a possibility. It is this latter problem that is difficult to resolve, as comparing life with defects to non-existence imposes a metaphysical difficulty (a similar problem is raised in legal discussions of euthanasia). The impossibility of measuring the difference between non-existence to life with disability, also leads to the difficulty or impossibility of measuring the damage, which normally has to be compared to a former undamaged situation.
3. The right not to be born: the assumption in wrongful life cases is that positive results of the genetic or medical tests would lead to abortion, i.e., to the non-existence of the subject of the right. This claim is based on the controversial argument that people have among their rights a right not to exist, not to be born. According to Heyd (1995), such a right is logically disputable. "For, if the fetus is a person having moral standing, than it is not clear what could be meant by the right not to be born, not to come into existence. And if it is not a person, how can it have any rights?" (p. 294). This, continues Heyd, is of course different from the right of parents to abort their future child ("wrongful birth"), which is conceptually coherent, though morally disputable.
4. Sanctity of life: wrongful life suits posit a symbolic danger for the sanctity of human life, as they refuse the claim that life in any form is more precious than non-life and hence, no life is unworthy of living. Thus social policy considerations in favor of preserving, protecting and encouraging human life, forbid

wrongful life compensations, which are feared to devalue the life of the handicapped. Dunne and Warren (1998) argue against the logic of wrongful life claims, as they refuse to understand genetic difference to be a genetic "defect", and to accept the right of mothers or parents not to carry what they perceive to be a "defective" individual. According to them, this right contradicts public policy, which is based on treating the disabled equally and respectfully. Moreover, they question the causation chain: if the parents had known about the positive test result, they would have chosen an abortion. Dunne and Warren suggest another option in which the parents might seek further information regarding the condition, which could result in their decision to bear the child and nurture it to the best of their abilities.

5. The slippery slope argument: wrongful life claims are seen as leading to the slippery slope problem, which is troubled by the question: where is the line between life with or without value drawn, and who is allowed to set the criteria? (Heyd, 1986). Accordingly, the slippery slope argument also fears the move from the right to be born without defects to the right to be born in a better condition, with a better genetic make-up, citizenship, racial or ethnic identity or even with better parents (Heyd, 1992). Hence, wrongful life suits are also feared to legitimize wrongful life suits of children against their parents for having them, whilst knowing their medical situation. This threatens the right of self-determination of the parents.

To conclude: the difference between wrongful life and wrongful birth suits is not in the type of claim but in the injured party. Practically, or compensation-wise, it is not a major difference. In fact, most such cases around the world are solved with the less problematic wrongful birth suit, as only little practical injustice is caused by denying the wrongful life action. This is because wrongful birth suits protect the child until majority and the survival of a severely defective child beyond that age is anyhow unlikely. As a matter of fact, wrongful life suits are more comprehensive only in the sense that they allow a child to sue his doctors in the case of his parents' death or as an adult. Consequently, the significant difference between the two claims is more philosophical than practical. A large moral-symbolic difference exists between compensating a child for its "wrongful life" and compensating his parents for losing the option to abort him/her, or for the extra costs of raising a disabled child (wrongful birth). This is especially true, since wrongful birth claims do not go as far as suggesting that certain kinds of life are wrongful. Therefore, my next step is to try and make sense of the symbolic difference between wrongful life and wrongful birth claims in different national contexts and to show how different legal systems have solved this problem.

GERMAN LEGAL SYSTEM'S POSITION

Wrongful life: the courts in Germany reject wrongful life claims of the child. The general opinion among German courts is that the unhealthy child cannot recover damages for its condition and likewise, his parents cannot recover damages on his behalf. In cases of alleged wrongful life, it is the German court's opinion that there can be no measure of damages (Harrer, 1994). The seminal case of the German courts is 86 BGHZ 240 (1983), in which a pregnant woman had rubella during early pregnancy. The attending surgeon negligently failed to diagnose the disease and as a result, the woman did not abort the affected fetus. As the child was born handicapped, the mother sued the surgeon and demanded compensation on both contract and torts theories. The court said there was no direct contractual relationship between the child and surgeon and therefore, it turned down contract theory. Furthermore, the tort claim was not accepted because there was no violation of a right of a child. The court emphasized that the surgeon did not cause the injury, but only prevented a possible abortion because of a wrongful diagnosis. The court stressed that if anything, the child was indebted to the surgeon for his or her life. Moreover, the court emphasized that it is not possible to compare the state of existence to non-existence and the state of being alive and handicapped to never being born. Life, even with disabilities cannot constitute damage, since human life is a legally protected object of the highest value and there is no "right of nonexistence". Moreover, if life with disability were viewed as a legally compensable injury, then such a finding would violate the interests of all physically and mentally disabled people (Harrer, 1994).

Wrongful Birth: the German Supreme Court has approved wrongful birth claims since 1980 (The first cases were 76 BGHZ 249 and 76 BGHZ 259, both decided on the same day) and since then it consistently holds that the birth of a physically or mentally handicapped child entitles the parents to full compensation for the total burden of the maintenance of the child (Harrer, 1994). The Supreme Court also holds that under special circumstances, such as the prolonged suffering of the child, compensation should not be time limited, since in these cases the parents are usually obliged to render services and maintenance for the whole of the disabled child's life. The restitution has to place the family in the same situation that would have existed if there had been no damaging event and in our case, if an abortion had been performed. The extent of protection afforded to the parents covers total maintenance and is not limited to the extra costs of being handicapped. Furthermore, the Supreme Court has ruled that the burden of maintenance should not be divided between that of a hypothetically healthy child and that of a disabled child (Harrer, 1994).

ISRAELI LEGAL SYSTEM'S POSITION

Wrongful life: such claims of the child have been rejected by the overwhelming majority of courts around the world. Thus, Israel is a world exception, as in the famous Zaitsov v. Katz, C.A 540/82, 40 P.D. (2) 85) verdict of its Supreme Court, which took place in 1986, it recognized this legal claim (Heyd, 2002). In this case a woman had turned to genetic counseling prior to her marriage, to check whether Hunter's disease – which existed in her family – could harm her future children. If so, she wanted to avoid having sons, as they are the risk group for this disease. The counselor advising her mistakenly told her the condition could not be inherited by her children. Later on, a child with Hunter's disease was born and the mother sued the counselor. In the District Court her claim was turned down but in an appeal to the Supreme Court, the cause of action was recognized by four out of five justices. The conceptual problems of such a law suit were solved as follows:

Normally, compensation to a plaintiff is based on the principle of restoring the injury to "*status quo ante*", i.e., to create through compensation conditions which resemble, as much as possible, life before the injury. When a plaintiff prefers non-existence to living with disability, s/he is claiming that the gap between the two situations is the damage suffered by her. However, can a court accept this claim in a "wrongful life" suit? In the Zeizov petition to the high court, two justices of the Israeli Supreme Court, did not shy from comparing non-existence to life with disability (Justices Miriam Ben-Porat and Dove Levin). They accepted that in certain, rare and extreme situations, a "reasonable person" could conclude that a person would be better off not born at all than brought into the world impaired. This position is obviously extreme, as it opens up the difficulty of deciding what problems are severe enough to allow for the wrongful life cause of action (Heyd, 2002).

An alternative approach, which seeks to avoid this logical dilemma, was preferred by two other justices of the Israeli Supreme Court (Aharon Barak and Shlomo Lewin). The opinion of these judges was that the physician's liability towards the child in the "wrongful life" situation rests on the plaintiff's right, once born alive, to life without defect brought about by professional malpractice. Thus, the genetic counselor's negligence is responsible not for the granting of life or prevention of non-life, but for causing impaired life. Consequently, the plaintiff's life should be compared to hypothetical life without harm. The defendant's professional negligence yielded at once two results: the child's birth and her inevitably handicapped existence. The justices claimed that although logically, these two results are intertwined, they could nevertheless be separated conceptually (Heyd, 2002).

According to an interview with lawyer Samuel Jellinek, author of the book *Wrongful Life: Rights of Claim and Compensation* (Jellinek, 1997), in reality, wrongful life suits are very rare in Israel and wrongful birth suits are also possible. Therefore, the interesting question is why the Israeli court has declared some types of life to be unworthy of living, and recognized this cause of action? The supporters of this stance, such as Jellinek, admit that wrongful life suits confront logic with justice, but argue that following logic, reasoning leads to imposing injustice on a suffering plaintiff whose claim is not recognized and who is not given a remedy for a wrong.

Shapira (1995) explains this stance as resulting from the Israeli abortion law. Thus he writes that *de jure* as well as *de facto*, a permissive stance on abortion for eugenic reasons like the one existing in Israel, actually provides a solid foundation for an impaired newborn's assertion that had it not been for the genetic counselor's negligence, his parents might indeed have opted for legal abortion. "*In this respect, legal recognition of the cause of action for "wrongful life" is rationally linked to the corresponding legal endorsement of parental authority to refrain from conceiving or giving birth to an impaired child*", writes Shapira (p. 25).

Yet, it is interesting to note that while new reproductive technologies are usually said to encourage physicians to confer to the embryo/fetus the status of a patient, this does not happen in Israel. Rather, in Israel the autonomy of the fetus is usually not recognized, as it is perceived as part of its mother and therefore, as we have seen in chapter five, the attitude towards abortions is permissive. Strangely though, when wrongful life suits are the case in point, the fetus' rights, now in the form of a born child, are recognized *post factum*, in contradiction to the general Israeli understanding of the relationship between the pregnant woman and her child. Said bluntly, it seems like the only right the Israeli fetus holds separately from its mother, is the right not to be born handicapped.

GENETIC COUNSELORS AND "WRONGFUL LIFE": THE PHILOSOPHICAL CONCEPT AND THE LEGAL THREAT

Counselors, who set the standard of care, may feel compelled to encourage women to choose genetic testing, both out of concern for the parents and future child and for fear of legal liability. The paradox is that the more the profession encourages genetic testing and parent's expectations, the more it is trapped into providing the services, due to the perceived threat of litigation (Charo and Rothenberg, 1994) and the more its practice of medicine becomes defensive. Yet, the threat to the counselor is not greater where wrongful life suits are recognized, since practically, wrongful birth suits are as threatening. Thus, counselors in both states were worried about law suits against them.

For example, a German M.D. specialist in human genetics, a man working in the private sector, said:

"There are no wrongful life suits here, but wrongful birth suits put tremendous pressure on us. We have tremendous insurances to pay, and because we fear legal actions against us, we talk about low risk possibilities and always mention that despite the tests, something can always go wrong, and the child may be born with a problem".

In Israel, at a genetic round-table meeting of Israeli counselors, which took place in October 2001, the need to write consensus statements was discussed and supported on the grounds of the need to be defended against legal actions and not so much due to ethical considerations. "*If we make a mistake, at least let's do it all together...*",the counselors bitterly joked. Furthermore, according to Cohen-Almagor and Snir (2000), who write about decision making in Israel's abortion high committees, the fear of litigation also affects abortion committee members. Thus, they mention Zeizov's verdict from 1986 to be influencing committees' decisions in the direction of accepting abortion appeals, due to the fear of getting involved in juridical confrontation, in case of refusal. Also in my interviews with the counselors in Israel, it was repeatedly mentioned that abortion is "safer" from the doctor's point of view. To solve this bias, some of the counselors were expressing the wish for someone to sue for liability in cases involving erroneous abortion of a presumably impaired, but actually healthy child. For example, an Israeli gynecologist and genetic counselor, a man who is in charge of a hospital genetic unit, said:

"In Israel we are very tolerant towards abortions and intolerant towards sick children. Part of the reasons behind this attitude, are very practical. Jews like to sue and we have to protect ourselves against this tendency. The picture will be reversed on the day someone sues for wrongful termination. Or, if there was a Church or a Kadi to protect us".

Another Israeli pediatrician and a genetic counselor, a man working in a hospital, said:

"The legal question no doubt affects our practice. If there is a doubt, why take a chance? No embryo that has been aborted has ever sued us. But born children can sue. So why should we get into this trouble?"

This claim, however, is simply false. A legal suit for an erroneous abortion has taken place in Israel and the parents were compensated.[1] Strangely enough, none of the counselors I interviewed was aware of this precedent, an amnesia that could result from the fact that it does not fit the cognitive/ideological schemata that Israeli counselors have concerning such matters, or from the fact compensation for an erroneous abortion is far cheaper than for "wrongful life".

The extra caution taken by the counselors has the side effect of raising counselors' and in turn patients' anxieties, as was mentioned by an Israeli pediatrician and the head of a hospital unit, a woman, who described her practice in light of the legal situation as follows:

"Legal suits frighten us. Not long ago only gynecologists were exposed to it, but today it is also us, the geneticists. Therefore, I explain myself very carefully, say everything a few times, let people sign and check again and again whether I have been understood. I also give a lot of what I consider extra information, about every additional test, even for very minimal risks".

A general practitioner and a genetic counselor, a woman who was less concerned with her own threat and more empathic to the patients' situation said:

"The problem with the legal situation is that it leads to over testing and to great anxiety among future parents".

Yet, although both Israeli and German counselors take precautions against legal action, the interviews reveal that Israeli counselors are more concerned than their German colleagues about legal action against them, and of course more prone to opt for an abortion. Once again, as in our former discussion of late selective abortions, this difference may not be explained by the actual legal threat (of "wrongful life" or "wrongful birth" suits), which is practically (though not symbolically) very similar, but rather, by the general understanding of the "wrongful life" debate in both societies.

When asked to react to the statement: "It is not fair to a child to bring it into the world with a serious genetic disorder", a statement saying that from the child's point of view, some kind of life is not worth living, the largest group among the German counselors disagreed (47.3% and 52.6%), as opposed to the majority of their Israeli counterparts, who agreed with this claim (71.4% and 66.7%) (see Table 1).

---

[1] Leah Dayan et al. v. Clalit Health Services, C.C. 1037/93, 9.12.98, D.C.

TABLE 1. It is Not Fair to a Child to Bring it into the World with a Serious Genetic Disorder

	1994–96				2000–01			
	Strongly Agree + Agree	Neither Agree nor Disagree	Disagree + Strongly Disagree		Strongly Agree + Agree	Neither Agree nor Disagree	Disagree + Strongly Disagree	
Germany	14.4%	38.3%	47.3%	100%	5.3%	42.1%	52.6%	100%
Israel	71.4%	9.5%	19%	100%	66.7%	20%	13.3%	100%

As Table 1 shows, German counselors strongly rejected the idea that impaired life is not worth living, while the majority of Israeli counselors, in both time periods, supported it. To put it in the harsh words of an Israeli MA genetic counselor, a woman working in a hospital:

"I don't see life as something so holy. If life is shit, it is not holy. I don't venerate life in the garbage". It seems that many Israeli counselors feel the same, as they largely accept the idea of a "wrongful life", in contrast to their German counterparts who reject it, but also feel that an open debate about "wrongful life" is totally blocked in contemporary Germany, due to its history (see Chapter 8).

THE POSITIONS OF DISABILITY ORGANIZATIONS IN ISRAEL AND GERMANY

A similar difference to the one existing between Israeli and German counselors in their positions towards the notion of "wrongful life" also exists between the positions of disability organizations in both societies. In Germany, disability organizations are actively involved in the public discussion concerning reproductive genetics, as they fear prenatal diagnosis, wrongful life claims and selective abortions devalue the life of their own kind and turn them into accidents that have not been prevented due to medical negligence (Erikson, 2003; German National Ethics Council, 2003). In contrast, in Israel, in all the years that I have been following this subject, I have never encountered any public statement by any Israeli disability organization that opposes these practices. Contacting some of the organizations, it was only in the Association for people with hearing disabilities that I was told they perceive their members to be a special community, with a different culture, not with a disability and thus selective abortions on the grounds of deafness are opposed by them. Yet, this was not done publicly and despite their ideological position, the Israeli Association for people with hearing disabilities is not taking any measures against selective abortions on the grounds of deafness, which do take place in Israel. Similarly, Raz (2004), who studied the views of leaders of disability rights organizations and support groups for people with genetic conditions in Israel, and compared them to the discourses of their North American counterparts, found that unlike the prevalent position of their American fellows, Israeli activists are generally in favor of prenatal diagnosis and selective abortions. However, these attitudes go hand-in-hand with their commitment to the support of disabled individuals. Raz explains this as an Israeli two-fold view of disability, which separates prenatal preventive testing from postnatal support of disability and attributes positive connotations to eugenics, which is understand to improve the health of the progeny. Another striking example of the fact that the Israeli disabled do not resent the idea of selective abortions of their own kind, or the idea of a wrongful life, or even of infanticide, is the story of Yoav Kreim, the present spokesmen of Israel's disabled persons organization. Kreim was first known to the Israeli public as a child, when his father, a known literary critic said in a newspaper interview, (in front of his then 12-year old handicapped boy, who sat next to him), that he thought it would have been better to let his child, suffering from CP, die when he was one year old. In an interview with Kreim junior, who was by the time of the second interview twenty-seven years old (Lori, 2003), he said he was not angry with his father, who wanted to save his child and the rest of the family the tremendous suffering involved in living with CP. On the contrary, Yoav Kreim said his father was right, since it demanded tremendous efforts to bring him up and it involved a high price for the entire family. Thus, because the handicapped are mainly their family's concern and not society's (a situation criticized by Kreim), parents should be given the opportunity to choose.

"*My father said that if he had known in due time that my life would be dependent on medical instruments, he would not have fought to save it. I was born prematurely. All that was necessary was to take me off the medical support. I don't understand this to be killing*"(my translation), said Kreim junior, who rather believes it to be his parents' moral right. No opposition from his fellow men was publicly heard after this interview.

In contrast, in Germany it is unthinkable that the head of a disability organization would express such opinions, as many German disability organizations are fighting very hard against PND and selective abortions. This non-existence of opposition to the practice of PND and to selective abortions coming from disability organizations in Israel, as opposed to German disability organizations' strong opposition, is clearly demonstrated in the German and Israeli geneticist's answers to the question "Do any of the following act as sources of opposition to genetics services in your country"?, which was presented to them in the Wertz and Fletcher international survey of 1994–96. (See Table 2)

TABLE 2. Opposition to Genetic Services

	Israel (% checked)	Germany (% checked)
Religious groups	68.2%	56.2%
Woman's groups	4.5%	61.8%
Political parties*	18.2%	28.3%
Advocates for Person with Disabilities	**4.5%**	**74.1%**

* *In Israel, the opposing political parties are religious parties.*

The position of many of Germany's disability organizations is best demonstrated by the story of cystic fibrosis (CF), the most common inherited genetic disease in Germany (carrier rate of about 1/30). Despite the fact CF is not an extremely rare medical condition in Germany and despite the fact that CF patients usually die in their 30s or 40s (or earlier) and their life involves a lot of medical attention (Vardimon and Ben-Rafael, 1999), no screening tests are offered to the German public, in accordance with the position of the CF organization. At an international workshop dedicated to Evaluating Cystic Fibrosis Carrier Screening Development in Northern Europe, which took place in 1992, the representative of the German CF Association, the late A. Kersting-Wilmsmeyer said on behalf of the association:

"...We see our disease as a challenge which may influence our lives in a positive way, for example towards a greater sensibility for the discriminated and towards a deeper sense of life. Many of us feel their life to be more conscious and desirable than their non-handicapped fellow men. .. Don't you think that living with trouble and knowing one's own limitations means having a much more intensive life than that of careless so-called non handicapped people? We therefore demand: Gene testing only if personally wanted, no advertising and as a precondition in any case a human genetic consultation!" (Kersting-Wilmsmeyer, A. "Views of the German Cystic Fibrosis Association" Quoted in Nippert et al., 1998, p. 31).

In contrast, CF screening is widely advertised and practiced in Israel, without any objections from CF patients and without any glorification of life with disability, a theme which constantly reappears in the German discourse on this issue. Thus, a very important theme that hovers over the question of "wrongful life" (in the discourses of both the disabled and the professionals) is obviously that of the meaning attributed to human suffering, both at the individual and at the communal level. While the German glorification of disability is often understood to be a counter reaction to the atrocities of the Nazi past, my suggestion is that it also echoes Christian religious understandings of the role of suffering. Likewise, Israelis' intolerance towards the disabled follows a Jewish logic, which sees no point in suffering. The religious background for such dissimilarities is explained by Steinberg (1999):

"Christianity views pain as the punishment for the original sin and hence, it is a welcomed experience and it should not be prevented and should be gracefully accepted. Judaism accepts the idea that pain might be a punishment for sins. However, it negates the notion of original sin and it certainly does not regard it as a virtue or sign of grace. Rather, pain and suffering is an unwanted curse, even if it has a purpose. Hence, it should be prevented and treated..." (Steinberg, 1999, pp. 80–81)

In addition, Zvi Marx (2002) states that attitudes to disability evinced in the Rabbinic and Halakhic literature, are often disturbing to contemporary sensibilities, as Jewish Halakhic culture is to a large extent a competence-oriented culture of "learners", which excludes the disabled, who are exempted from the obligation to fulfill many precepts, an exemption which leads to their devaluation and stigmatization.

In accordance with the existing (in Germany) or non-existing (in Israel) political struggle of disability organizations, Israeli genetic counselors feel less obliged to involve organizations of people with disabilities in their policy decisions. Moreover, when I mentioned to Israeli counselors that in Germany, reproductive genetics is sometimes seen as a contemporary form of racism against the disabled, a pediatrician and a genetic counselor, a man working in one of Israel's largest hospitals and representing the most extreme Israeli attitude towards the disabled said:

"I think that saying we are practicing a form of racism is cheap demagogy. The deaf, for example, wish they were not deaf, but they say: I was screwed, so I can now screw others. It is like AIDS patients who infect others, saying otherwise it is racism against them. It is a very hypocritical way of looking at things. Maybe some "professional dwarfs" and deaf people want political power, and that's why they want more handicapped people to be born. But I don't have to agree with it".

Regarding the same theme, a German human geneticist and a hospital genetic unit manager, who took a critical attitude, told me:

"I don't think society should pressure people to have disabled children, I believe it to be a personal choice. But here in Germany, the disability organizations see it differently. I believe the parents of the disabled are not so different in Germany than elsewhere, privately they don't want another sick child, but they don't dare argue in public. The ones who argue are the talented and the smart among the handicapped, those with political power. The thing is that only they are heard, not the really miserable ones".

The dramatic difference in the tone of these two similar opinions and the fact that the German counselor feels that in his country no free and open discussion of those themes is possible, will be further explained in the next chapter, which is dedicated to the question of the "good society" and its genetic composition and to the communal meaning of suffering, as they are understood in both societies.

CONCLUSION

This chapter started out by examining legal responses to the notion of "wrongful life" in Israel and Germany. In accordance with my general argument, it once again confirmed that the general concept of "life", in terms of the conditions under which a certain form of existence is perceived by society to be (un)worthy of protection or (un)worthy of living, are very different in the two studied societies. According to the German legal system, no "life" can be declared unworthy of living, as opposed to the Israeli situation, in which wrongful life suits are recognized. Turning to the genetic counselors, the picture becomes somewhat more complex. Reprogenetics is a far more controversial practice in Germany than in Israel. Therefore, German counselors are exposed to many objections to their practice, whereas such criticism hardly exists in Israel. Consequently, German counselors more often than their Israeli counterparts feel pressured to defend their expertise, which is under attack. This was clearly shown in this chapter by some of the German counselors' attitudes toward the positions of disability organizations. Similarly, among the Israeli counselors, opinions regarding those very delicate matters were not uniform, and it is important to note that some Israeli counselors were clearly more critical than others regarding the prevalent understandings concerning the problem of a life (un)worthy of living in their society. Yet, they never felt attacked by their community for practicing genetic counseling.

Nevertheless, generally speaking, the findings of this chapter suggest that in both states, the professionals seem to accept the typical logics of justification found in their respective societies concerning the moral dilemmas associated with the concept of a life (un)worthy of living, as far more Israeli than German counselors believe some forms of life to be unworthy of living. Thus, the legal situation combined with the counselor's acceptance of its logic, makes Israeli counselors more prone to encourage selective abortions, despite the fact that the actual legal threat counselors in both states are exposed to is quite similar, due to the existence of wrongful birth claims.

Likewise, in both countries, disability organizations' positions are in agreement with the general understanding concerning the prevalent notion of a life (un)worthy of living in their society. German organizations reject the idea of "wrongful life" and its derivatives, namely genetic tests and selective abortions, as they tend to glorify life with disability and to fear that selective abortions devalue their own lives. Alternatively, Israeli disability organizations pose no opposition to the practice of reproductive genetics and its consequences. Rather, they differentiate between the rights of the living disabled and the prenatal rights of disabled embryos (Raz, 2004), and accept the idea of non-existence as preferable to some forms of life.

CHAPTER 8

THE CONFLICTS BETWEEN INDIVIDUALS, FAMILIES AND SOCIETY, AS WELL AS BETWEEN DIFFERENT FAMILY MEMBERS, EMBODIED IN REPRODUCTIVE GENETICS

The question of the conflict between the interests of individuals and the supposed interests of society embodied in reproductive genetics, which is as old as the eugenic movement itself, can be formulated in many ways: does the public have a legitimate interest in individuals' reproductive choices? Or alternatively, should parents have complete autonomy regarding their reproductive decisions? And if so, how should society treat individual decisions that are costly to it? Furthermore, are there only two parties involved in this conflict (parents and society) or do unborn fetuses have their own rights, and if so, who represents them?

Concerning this conflict between parents-fetus, parents-society and fetus-society, Heyd (1995) writes that the liberal ethos typically gives priority to the parents' rights, since they are actual rather than potential human beings and since they are identifiable individuals, rather than vaguely defined collective entities. However, the supposed interests of society challenge this priority, in the name of the individual (liberal) rights of the fetus and in the name of allegedly weightier social interests.

But what does "weightier social interests" mean? Answering this question, which is the aim of this section, is not a simple task. That is due to the fact that the term "social interests" is in itself a black box, used by different and even opposing parties, to justify their positions. In the case of the comparison between Israeli and German cultural understandings of the social interests embodied in reproductive genetics, opening this box reveals antithetical understandings of the very stuff those interests are made of.

This is surprisingly true despite the fact that the guiding principle of present-day reproductive genetics, or in a more critical phrasing, present-day "voluntary eugenics" (Wertz, 1998), is parental autonomy, a principle which helps to draw a line between the atrocities of the "old" eugenics, which was enforced by the state, and the repro-genetic practices of today. Due to this prevalent ideology, in both Israel and Germany, the vast majority of counselors contend that the aim of reproductive genetics should be to enhance parents' autonomy, and not to serve any eugenic social ideology. Yet, among the Israeli counselors, quite a large minority group exists, that does not shy away from explicit eugenic targets. Such a group does not exist in Germany as is shown in Table 1:

The fact that German counselors disconnect themselves so strongly from eugenic practice and ideology is best explained by the German past, in which eugenics was such an important element of the Nazi regime. Accordingly, the major moral lesson

TABLE 1. An Important Goal of Genetic Counseling is to Reduce the Number of Deleterious Genes in the Population

	1994–96				2000–01			
	Strongly Agree + Agree	Neither Agree nor Disagree	Disagree + Strongly Disagree		Strongly Agree + Agree	Neither Agree nor Disagree	Disagree + Strongly Disagree	
Germany	1.6%	3.8%	94.5%	100%	–	3.6%	96.4%	100%
Israel	14.3%	–	85.7%	100%	13.3%	13.3%	73.3%	100%

learned from this totalitarian, murderous past is that the state should not interfere with private affairs. Yet, things are not so simple. Looking deeper into the nowadays common German understanding of this matter reveals that German history did not only lead to the conclusion that reproductive decisions should be completely individual. Rather, a reverse moral understanding of society's good with regard to the genetic makeup of the population, according to which the "good society" must be genetically varied, has been developed, thus putting society's interests (once again?) before individual ones.[1]

Therefore, moving to less straightforward eugenic statements, the difference between Israeli and German counselors, in their perception of the "social good", deepens (See Tables 2 and 3), as the Israelis are revealed to support a more eugenic attitude, as opposed to their German colleagues, who believe the "good society" must now include the disabled, who were once so violently persecuted.

Being asked about whether the existence of people with severe disabilities makes society more rich and varied (Table 2), which is a statement giving positive value to disability, from society's point of view, the German and Israeli counselors expressed

TABLE 2. The Existence of Persons with Severe Disabilities Makes Society More Rich and Varied

	1994–96				2000–01			
	Strongly Agree + Agree	Neither Agree nor Disagree	Disagree + Strongly Disagree		Strongly Agree + Agree	Neither Agree nor Disagree	Disagree + Strongly Disagree	
Germany	42.4%	40.0%	17.7%	100%	61.4%	33.3%	5.3%	100%
Israel	10%	30%	60%	100%	40%	33.3%	26.9%	100%

[1] In a similar vein, writing about the embryological research debate in England and Germany, Richardt (2003) writes that "In contrast to Great Britain, where the effects of the legislation were judged on the basis of the effects on an individual person, the common good or the interests of society as a whole were used as the basis for the rejection of embryological research in Germany" (p. 110).

TABLE 3. It is Socially Irresponsible Knowingly to Bring an Infant with a Serious Genetic Disorder into the World in an Era of Prenatal Diagnosis

	1994–96				2000–01			
	Strongly Agree + Agree	Neither Agree nor Disagree	Disagree + Strongly Disagree		Strongly Agree + Agree	Neither Agree nor Disagree	Disagree + Strongly Disagree	
Germany	9.3%	–	90.7%	100%	3.6%	8.8%	87.8%	100%
Israel	66.7%	14.3%	19%	100%	24.1%	41.4%	34.4%	100%

different positions, as 42.4% in 1994–96 and 61.4% in 2000–01 of the German counselors agreed with this statement, while only 10 and 40% (respectively) of their Israeli colleagues agreed with it. In 1994–96 about 3 times more Israeli than German counselors disagreed with this claim (60 and 17.7%), and in 2000–01 about 5 times more Israelis disagreed with this statement (26.9 compared to 5.3%).

Furthermore, when asked to comment on whether it is socially irresponsible to knowingly bring an infant with a serious genetic disorder into the world in an era of prenatal diagnosis, Israeli and German counselors expressed the following positions: (Table 3)

Percentages of agreement with the idea that nowadays it is **socially** irresponsible to knowingly give birth to an infant with a serious genetic disorder were 6-7 times higher among the Israeli group (66.7 compared to 9.3% in 1994–96 and 24.1 compared to 3.6% in 2000–01). The percentages of disagreement were 4.7 times higher among the German group in 1994–96, (90.7 compared 19%) and 2.5 times higher (87.8 compared to 34.4%) in 2000–01. However, the passing of time has clearly toned down Israeli counselors' eugenic attitudes, as was discussed in chapter four, presumably due to the entrance into the profession of a large group of younger counselors who feel less obliged to serve the state, in what may be part of a general Israeli tendency towards more individuation.

What are the logics of justification behind the diverse Israeli and German positions?

Fiedler (1996) writes:"*...deep in the mind of all of us there persists a desire to murder the disabled, even when they are infants, and even as we weep for them*"(p. 43). Thus, as a means of handling our murderous, pity and fearful emotions, handicapped people tend to be stereotyped, either negatively or positively, but in any case as something more or less than "human". In the German case, in contrast to the past, positive stereotyping now prevails, as handicaps are considered a very crucial part of society, serving an important moral task, that of making the rest of society more human. This position is very clearly articulated in the writings of two prominent German geneticists, Schroeder–Kurth, T.M and Huebener J. (1989) about ethics and medical genetics:

"A basic problem is what kind of mentality the growing possibilities of genetic counseling and prenatal diagnosis are creating among the public. There is the danger that a "biological norm" may be created, which classifies genetically burdened human beings as "avoidable" and consequently unacceptable by society. This would create a "duty to have a normal (non-handicapped) child". This position no longer recognizes the "diakonic" task of the handicapped in society. By "diakonic task" we mean that the handicapped, by their presence, perform a real social service. Diakonic comes from the Greek word for servant, and is used in the New Testament to describe the Order of Deacons who cared for the poor, sick and elderly in the early Christian Church. The handicapped are deacons in a symbolic sense. Their presence serves to make us reflect on what it means to be human. Their presence also serves to draw communities together in making sacrifices to provide for their wellbeing, and thereby enriches the fabric of human relationships. A society without the handicapped would lose its sense of community and its willingness to care for others. The ideal of perfect health sacrifices the reality and complexity of life and human relations, which do not exist without suffering" (p. 169).

In their writings, Schroeder–Kurth and Huebner give a positive value to disability, not because it is good for the individual handicapped, but rather because it serves society and one may even say, Christian society. In that sense, this contemporary social moral understanding of disability resembles the attitude towards lepers in medieval times, which were simultaneously negatively and positively stereotyped. In those times, according to the church, leprosy was interpreted as a holy disease and its victims were believed to be chosen by God, in order to grant them salvation in the afterlife. Their suffering was understood to be similar to the suffering of Jesus Christ, who suffered for all human kind, and Jesus himself was sometimes believed to reappear as a leper. As a result of this understanding of leprosy, the ill were treated by the Christian community with mercy and compassion, and charity for lepers was believed to provide forgiveness for the sins of this world (Navon, 1991). However, side by side with the described attitude, the lepers were also believed to be sinners who were being punished for their sins and accordingly, they were often excluded from their own communities.

In Germany today the moral-social status of the genetically abnormal is characterized by positive stereotyping, as their suffering is glorified (a common theme also in the discourse of the German disabled about their own lives, as was discussed in chapter 7) and as they are believed to serve society. Such an attitude repeatedly appears in the writings of Christian religious figures discussing the challenges of reproductive genetics in Germany. For example, Johannes Gruendel (1987), a Catholic moral theologian from the University of Munich, writing about prenatal diagnosis, argues that in order to prevent the formation of a society which values only healthy life, one should stress the value of every human being. Furthermore, acceptance of the handicapped should be promoted, as they should not be devalued but rather seen as a hint and reminder of the suffering of Jesus Christ. Similarly, Zimmermann and Zimmermann (2000) in their article entitled "Is there a right to a healthy child?" link reproductive genetics with a hedonistic world view, which aims to reduce suffering and increase happiness at the cost of neglecting the fetus' right to life. Similarly, a statement issued by the Council of the Protestant Church in Germany (2001), fears that therapies having to do with the human embryo nourish the illusion of a world free of suffering.

In contrast, the attitudes to disability evinced in the Jewish Rabbinic and Halakhic literature often devalue and stigmatize the disabled (Marx, 2002), who have no function concerning the morality of the community. However, it is worth noting that despite this general Jewish attitude, a lot of organizations that support the disabled in contemporary Israeli society are run by the religious community.

One way or the other, it is important to acknowledge that just like negative stereotyping, positive stereotyping also does not understand disability to be a non-stereotyped human variant. Rather, both positions perceive the disabled as more or less than human and thus, the "good society" must either exclude them or cherish and admire them, for society's own wellbeing.

Positive stereotyping was common in the positions expressed by some of the German counselors in this study, who believed genetic diversity to be socially, morally and also biologically important. For example, a gynecologist, working in the private sector said:

"Being around people with disabilities, you get a stronger sensitivity for the fact that your own life can also change from health to illness, or poverty, from one minute to the other. The disabled remind us of that. They are a strong reminder of real life, outside of the TV screen, where not everybody is young, healthy, rich and funny. They also stop us from becoming more and more uniform, which is dangerous from an evolutionary perspective".

Another counselor, a female Gynecologist working in the private sector, in cooperation with an IVF clinic said:

"We only look for money and for a good job and forget all the rest. Disabled children make you confront the other important things in life. It reminds us that there are people around us who need our help".

Whereas all German counselors paid respect to the idea of human genetic diversity, either for the sake of a humanistic ethos, or for the sake of biological diversity, Israelis completely mocked those ideas (See Table 2). Likewise, whereas German counselors felt their society had to be reminded of the fact that life is not all about money and fun, Israeli counselors did not share this feeling. For example, an Israeli women genetic counselor said:

"Do people with severe disabilities make society more rich and varied? Maybe on the philosophical level. But in reality, I want to see you spend one day with a child with CP or mental retardation, and then you will see what it really means. Then I will ask you again what you think of the high minded who say it's nice. There are enough things that make our lives diversified without suffering. One should not have disabled children in order to enrich the world or to make his neighbor more sensitive".

In a similar vein, an MA genetic counselor, an Israeli woman working in hospital, remarked concerning the idea of diversity as follows:

"I think the idea of diversity is bullshit. It can be supported only by high-minded people who don't really know the situation. It is easy to support until it happens in your own family. You can't say to people with a sick child that he makes society richer, despite the fact that sometimes the parents of the disabled try to glorify themselves and their children, as a means for coping with this terrible situation. As if having a disabled child is some kind of a reward. But it is all about rationalization, since no one would freely choose a sick or a retarded or a problematic child. And really, why should anyone care about society's diversity, when it comes to his own private suffering?"

Her colleague, a woman counselor thought the idea of diversity total absurd:

"Diversity as a cause sounds absurd to me. Should society enjoy itself at the expense of the suffering child or family? Such a claim does not convince me. Diversity should not be a goal. Anyway, the healthy are varied enough, and from a genetic or evolutionary point of view, the effects of our practice are completely negligible".

Being aware of the logical circularity of her arguments, an Israeli M.D, a specialist in human genetics and the former head of a large hospital genetic unit, accused society of being intolerant towards the disabled and so justified her resistance to the idea of having disabled children in order to enrich society.

"In an ideal world, where life has a higher value and where there is tolerance towards the disabled, I might have said yes, the disabled enrich society. Not in the sense of diversity, but in exposing society towards the less lucky and by this, increasing tolerance. Theoretically, society could be enriched by interaction with the abnormal, but practically, the disabled suffer from society's attitude towards them. That's why I don't think children should suffer to correct society's ways. Besides, I think its better that society becomes less diversified, and everybody will be normal, if it means less suffering. We are so diversified without the illnesses that I think that speaking in theoretical and general terms about diversity without thinking of human suffering, is simple luxury".

In Germany some of the counselors also criticized the idea that the existence of the disabled is important for society's morality. Whilst not mocking the idea of diversity altogether, they did point out that this ideology disregards the suffering of the child and the parents. For example, a German human geneticist and a hospital genetic unit director said:

"I do believe society is richer with the disabled. But who has to pay the check for the education society needs? At the end of the day it is the disabled themselves, and their families, who have to deal with all the problems, and I don't see it as their duty to live in order to educate us. I don't think my patients should suffer, for me to have an educational humanistic experience".

Other German counselors, who were unhappy about the common German understanding of the social role of their patients, attributed the current German trend to the German history and its enforcement of a eugenic policy, which still shapes the current German debate about the "good society" and the value of life. In the eyes of these counselors, the influence of the past is blocking open and honest moral debate.

For instance, a German general practitioner with a specialty in medical genetics, a woman working in a private office attached to a lab, expressed the following views:

"We Germans always want to be better than others, especially the older generation, which suffers from guilt feelings. It is a psychodynamic process. Therefore, we say the disabled are necessary for society. I partly agree with it, because none of us stays young and healthy forever and we had better learn how to deal with it. I also know a lot of handicapped people who don't mind us, so why should we mind them? However, when a couple has a sick child who can't sit, walk or talk, it is just awful. Yet, society does not want to hear about it, only about how important it is to have sick people among us. And then the Nazi theme pops up again, with all the troubles the disabled faced then. Yet, what I think is that by looking at the past, we just close our eyes to our contemporary reality. And so, everybody has to support the official opinion, and hide what he really thinks, or he may be strongly criticized".

Another example for such a position which criticizes the present effects of the Nazi past is that of a German human geneticist and a pediatrician working at the university (mostly in research) who, remarking about the notion of "wrongful life", said;

"Here in Germany, I cannot use the term "wrongful life". It is unacceptable, since it is Nazi terminology. Our history, in which we killed sick children, is so problematic, that it blocks any moral debate. Nazi accusations are raised very often. For example, when I opposed the embryo protection law, I was accused of being the contemporary Mengele. It was a very difficult experience for me. In this country those issues are very emotional, which is a big problem for an open debate".

A female pediatrician and counselor, working in a private office, also felt her country's past to be setting her unfair limits, professionally and privately:

"Our history is just terrible. We killed sick children. And that leads for example to the prohibition on euthanasia, unlike in our neighboring Holland. I would like to have the chance to die when I am old and suffering, but it is not possible here because it brings up the Nazi issue. It is 2001 now, its obvious that we have been horrible, and that science was terribly misused, but on the other hand I think we should open up and discuss the problems of today and not always those of the past, either when they concern the beginning of life, or the end of it".

Despite the fact some of the German counselors felt the burden of their nation's past to be too heavy, none of the German counselors I interviewed went as far as agreeing with the idea that it is better for society to be free of handicaps. Furthermore, the large majority of German counselors resent any cost-benefit calculations concerning prenatal genetics (see chapter 3), both because of moral justifications and because they believe their state to be rich enough to support the less privileged.

These positions are not shared by the majority of their Israeli colleagues, who despite a surface liberal ideology which glorifies parents' autonomy, find the decision to knowingly bring into the world children with genetic defects as socially irresponsible, and understand the "good society" to be one free of the genetically abnormal.

With regard to a question about the possible conflict between individuals and society embodied in prenatal genetics, an MA Israeli genetic counselor, a woman said:

"A family which chooses to give birth to a baby with Down's syndrome places the burden upon society. My own opinions concerning such a decision are contradictory. On the one hand I find the parents' decision to be just. But, later on I think they cannot turn to society and demand its support, as if it is society's problem, as if the life belongs to the individual but the bill has to be paid by the community. Yet, I am aware that a human society should care for the needy. That's the contradiction. I also understand religious people who don't abort, but our belief system is totally different. I am for knowledge and not for beliefs, and those are two parallel lines of thought that will never meet. That's why I am not sure why people with my own faith should support people with a religious faith. It's like with the orthodox Jews or with the Bedouins. They choose to have an endless number of children, but then they complain about their poverty. I think people should decide any way they like, but then they must also take the responsibility for their decisions".

Concerning the conflict between individuals and society embodied in her practice an MA Israeli female counselor said:

"The focus of my counseling is on the individual patient or the family. Yet, of course I have society at the back of my mind. First of all I convince mothers to abort because I know how hard it will be for them to raise a disabled child. But then I also think of society and about the burden for the state. I can't ignore the fact that finally it falls back on the state and its budget. If I know that people choose to carry to term an effected pregnancy, with a child who will end up in a state institution, I don't like it, but I don't say it this way to the parents. Yet, I do think the resources invested in this child could have been used for a better cause".

Likewise her colleague, a female counselor from a hospital in the Tel-Aviv area, argued that having a sick child is not fair to society, as it demands large resources that are needed elsewhere:

"In a certain way to knowingly have a disabled child is unfair to society, since it demands very large resources that could have been used otherwise. It's unfair when the situation could have been prevented, but was not. We gave them an opportunity for prevention but the couple didn't use it and then they demand support for the child's entire life, or worse, they just abandon it".

To summarize, Israeli counselors are far more concerned than their German colleagues about the burden the disabled place upon society and far less supportive of the idea that the disabled contribute to society.

Related to the question of society's duties towards the disabled and *vice versa*, is the question of parents' accountability towards their children, regarding their reproductive decisions, which is the theme of the next section.

PHILOSOPHICAL TRADITIONS AND LOCAL UNDERSTANDINGS OF FAMILY ETHICS: IS THERE A CONFLICT BETWEEN THE CHILD TO BE AND HER FAMILY?

The German Kantian philosophical tradition is an important influence on the moral reasoning surrounding reproductive genetics in Germany. As write Krones and Richter (2004), categorical argumentations around the term "human dignity" following from the Kantian maxim that human beings should be treated as an end and never as a means, are predominant in Germany. Such arguments are often found in discussions concerning the ideal relationship between generations, and the threats new reproductive technologies pose to them. Thus, German ethical debates concerning NRTs ideally view children as individual subjects, namely as ends in themselves and not as part of a social group, that is to say, their family. As a result, it is commonly argued that children should not serve the wishes or even the needs of their family, which in return should never be allowed to "design" them.

This line of ethical reasoning concerning NRTs is typical of the moral philosophy of Jürgen Habermas, the famous German philosopher and social critic, who has written extensively on the issue of reproductive genetics, and whose views became very well known. In his writings, Habermas (2001) fears reproductive genetics will lead to the slippery slope toward instrumentalization and objectification of humanity, as the eugenically programmed human being about whom he speculates, might end up deprived of the potential for a fully ethical existence of a self. In order to prevent this from happening, Habermas argues that parents should not be allowed to become the designers of their product children, who will thus lose their ability to choose whether they wish to live up to their parents' plans and expectations, or not. Habermas thus thinks of reproductive genetics as leading to an irreversible distortion of the generational relationship (Habermas, 2003; Prainsack, 2006b). Underlying his argument is the implicit assumption that social enhancement (through providing good education, intellectual stimuli, and a loving home to children) is morally acceptable and even desired, while genetic enhancement is not (Prainsack Barbara, personal communication).

A similar position was held in 1987 by the German government enquiry on gene therapy, whose argument's concerning this issue were as follows:

"Parents have no right to treat children as products or to program them according to their wishes. Human individuality and diversity is a product of nature and fortune. It is not for parents to design their children and in this way limit their individual freedom. Doing so would be to violate their integrity. Thus, for parents to seek to design their children according to their own wishes would be to usurp undue power over future generations; it is not for earlier generations to become the masters over future generations or for the dead to rule over the living". (Government Commission of Inquiry: The Opportunities and Risks of Gene Technology: Documentation of Reports Presented to the German Bundestag. Quoted in Sutton, 1996. p. 42).

This concern corresponds with the belief that parents are expected to accept their children "warts and all", an ideal which was clearly formulated in a famous speech by the former German President, Johannes Rau, concerning medical genetic technology. In this speech named "Will everything turn out well? For progress befiting humanity" (2001), Rau used the metaphor of a gift to describe what are the rights of parents with regard to their children. Describing children as gifts implies that they should not be chosen or picked, but rather accepted "as is". Furthermore, Rau argued that there is no right to healthy children or to children at all. However, children do have a right to caring parents and, above all, to be born and loved for their own sakes.

In contrast, different scholars studying the experience of pregnancy in the Israeli society have argued that the "ideal" Israeli (mainly Jewish) mother is expected to express her love and care for her future child by performing maximal diagnostic tests during pregnancy. According to Remennick (2006) "good motherhood" is understood by Israeli women to involve a "genetic responsibility" for the future offspring and other family members, namely the responsibility to prevent their future suffering. Similarly, Ivry (2004) has claimed that the Israeli pregnant woman is expected to wait and see whether the fetus is "worthy" of keeping, since selection is understood in the Israeli-Jewish society to be an expression of care and responsibility towards offspring.

The current German and Israeli ethical discussions regarding PGD (pre-implantation genetic diagnosis), and specifically regarding PGD for sibling donors, serve as a good example for the differences discussed above concerning the meaning of good motherhood and family obligations. The innovative technique of PGD may be used for different purposes, mainly for avoiding the birth of a child with unwanted characteristics, either in the form of a genetic "disease", or even in the form of its future sex. Recently, PGD was also used in order to select a future sibling donor, whose tissue could be transplanted after birth to a sick sibling who will be assisted, or even saved, by this procedure. PGD is extensively debated in Germany and at present prohibited. I argue that this prohibition is partly due to the fact that the moral reasoning prevalent in Germany concerning such matters, stems from philosophical abstractions about the meaning of human dignity and about how it is related to family relations and obligations. In German discussions of NRTs and the generational relationship, parents are often not trusted to do what is best for their child, but are rather suspected of violating its status as a subject. Correspondingly, opponents of PGD in Germany often do not think of the family in terms of a unified group with a common good, but rather in terms of different individual parties with competing interests, which may lead to violation of individual rights. The opinion of the opponents of PGD in the German National Ethics Council (2003), appointed to discuss reproductive genetics, serves as a good example for this kind of moral reasoning:

"Vis-à-vis their parents, children assume the position of subjects, and therefore must not be seen as objects of parental reproductive decisions... The demand to accept a child only on conditions laid down by the parents themselves and unilaterally to limit the responsibility for his existence cannot form part of the reproductive autonomy of the parents even if directed towards the presumed benefit of the child-to-be. This would be incompatible with the child's status as a subject". (pp. 70–71).

However, the conflicting interests of family members do not end here, as the opponents of PGD in Germany fear that allowing PGD for sibling donors might put pressure on the mother to become pregnant and have another child in order to save her sick child, as well as that the child brought into life after PGD may feel that she exists not for her own sake, but only in order to serve as a donor for her sick sibling. This logic of argumentation leads two members of the group of opponents of PGD in the German National Ethics Council (2003) to reach the conclusion that renunciation of parenthood should be the appropriate decision in cases of conflict.

This kind of a solution which involves refraining from parenthood is unheard of in the Israeli pro-natalist context that while stressing the utmost importance of having children, does not condemn parents' wishes to influence the genetic make-up of their future children, require a total acceptance of any future child "as is", or imagines family members to be individuals under constant threat of conflict. Rather, turning to the typical Israeli logics of justification regarding parents and children's rights, the picture turns around. As has been accurately marked by Ruth Landau (1996), having children is understood as a basic right in the Israeli society, which overrides even the right to having parents (thus, single motherhood as well as having children from sperm retrieved post-mortem are supported by the state and its medical establishment). Moreover, parents' needs and desires are not necessarily understood to endanger the child's status as a subject, as the relationship between an individual child and her parents and siblings is not imagined in terms of a conflict. Alternatively, as marked above, responsible parenthood is understood to involve selection or even designing of future children. The support for designing future children is best demonstrated by the Israeli ethical discourse and official stance regarding PGD for sibling donors, which is understood to hold a blessing for the entire family and not as a disputable medical practice. To give an example; at a special meeting of the Israeli Parliament's Science and Technology Committee appointed to discuss PGD, the issue of sibling donors was presented as a "life-saving" technology, which can thus not be morally questionable. The chairman of the committee opened the session by saying "*At today's meeting we will discuss another purpose of PGD, an indisputable one – saving life. A debate arises when things are not clear, but when saving life is the issue, there is no argument. ... For that, we have invited Sharon Harari, the mother of a child born with a rare disease. In an innovative procedure of in vitro fertilization and embryo selection, she conceived a baby for the sake of donating bone-marrow to her sick brother*". (The Science and Technology Committee, 2005).

The general support for PGD for sibling donors presented at this meeting was justified on the grounds of the experience of an Israeli mother who used PGD in order to save her child suffering from Fanconi's anemia, a rare genetic disease. Thus, a personal lived experience and not abstract moral principles concerning human dignity and the rights of future parents, served as the departure point for the ethical debate in the Israeli parliament regarding the use of PGD in order to select a sibling donor. In this discussion the purpose of saving life by designing life was understood to override any other ethical considerations, and the possibility of family conflicts that might result from this innovative procedure were not mentioned at all in the meeting. Rather, the procedure was presented as beneficial for the entire family. Similarly, the Israeli National Bioethics Council's guidelines concerning PGD state with respect to sibling donors that "*the future child benefits from being born to a family with no sick sibling or parent and the mother benefits since she avoids the mental distress of a dying child, making PGD a double blessing, for it brings and saves life simultaneously*" (Israel National Bioethics Council, 2005). Thus, it is the common interests rather than the potential conflicts which are highlighted in the Israeli discourse concerning PGD, and it's designing of future children.

Lately, even the especially controversial use of PGD for sex selection (for non-medical reasons) was allowed in Israel in rare cases, following a couples' success in convincing a committee appointed by the health ministry of the procedure's importance for their wellbeing (Yasur-Beit-Or, Meital, 2006). Thus, also in this case of sex selection, "designing" future children was not understood as an automatic moral wrong by the Israeli authorities. Rather, the controversial procedure was measured against the interests of the future parents, interests that were not automatically understood to contradict the rights of the future child.

It may thus be concluded that in contrast to the German dominant ethical outlook, in Israel it is considered far more legitimate for parents to "design" or select their children, as part of their parental responsibilities for the future child and its family, since the interests of different family members are mainly imagined to be in concert. As a result, parents who do so are not understood to be exercising their autonomy in an egoistic or immoral manner. To summarize, very much like in the case of the relationship between mothers and their fetuses (discussed in chapter 5), the Israeli discourse on NRTs and the generational relationship views children first and foremost as parts of their families and thus overlooks potential conflicts between different family members. Alternatively, the German discourse regarding families and their future children dominantly views children as separate, individual subjects and not as first and foremost connected to their closest relatives. Accordingly, the German discourse focuses on potential family conflict, which may result from influencing the genetic make-up of future children, while the Israeli discourse focuses on potential collective benefits.

COUNSELORS' VIEWS CONCERNING PARENTS WHO CHOOSE TO GIVE BIRTH TO CHILDREN WITH DISABILITIES

Returning to the counselors' view of parents' reproductive decisions, the binary stereotyping of the handicapped as more vs. less than human, or as a crucial element vs. a burden to society, is mirrored in the genetic counselors' feelings and attitudes toward parents who choose to give birth to children with disabilities. Such parents are perceived either as moral and psychological altruistic heroes or as "backward and egoistic fatalists", according to their national ethnicity. My interview materials demonstrate that more German than Israeli counselors express appreciation toward parents who decline selective abortions, put their trust in faith or God (and not in science), and accept their children "as is". Hence, a typical "German answer" to my question: what do you think of parents who choose to continue problematic pregnancies? Was given to me by a German pediatrician and a specialist in human genetics, a man in his mid-thirties, who said:

"I kind of admire parents who choose to continue a problematic pregnancy. Because my work makes me see all the scary things that can happen, I admire their faith and trust, which are quite rare those days. I see it also as a good sign for the parents' partnership that they agree to take the risk together".

German counselors do not often criticize parents who choose to terminate what they believe to be problematic pregnancies, as that would contradict with their expertise. However, such parents are perceived by the noisy opponents of selective abortions in Germany as egoistic and even as hedonistic. For example, Zimmerman and Zimmerman (2000), two German theologians writing for a German medical journal on the subject of the right to have a healthy child, argue against selective abortions and wrongful life suits on the basis of what they perceive as the hedonism of the parents. According to them, the aims of reducing suffering and increasing happiness cannot be balanced against the fetus' right to life. In addition, women who opt for selective abortions are often accused by German feminists of being influenced by distorted rigorous concepts of normality (Krones Tanja et al., 2006).

Dissimilarly, while a minority of the interviewed Israeli counselors believed parents who chose to continue "problematic" pregnancies to be courageous, the majority of opinions among Israeli counselors concerning such parents were negative. Whereas in Germany parents who opt for PND and selective abortion are often described in the public discourse as egoistic, it is rather parents who choose to continue problematic pregnancies that are perceived by some Israeli counselors as egoistic. Answering my question about her feelings concerning parents who choose to continue "problematic" pregnancies, an Israeli M.D. counselor, heading a hospital genetic unit said:

"I think that giving birth to a sick child epitomizes egoism. I believe that if this child could decide on his own, he would choose not to be born. ... Therefore I don't think it is fair to the child to suffer. It is also unfair towards his siblings. It is true that they might gain some maturity and sensitivity, but their emotional burden will be very heavy, especially since the parents don't live forever."

This difference in what an egoistic act means in the public discourse in Germany and the professional discourse in Israel, may be explained by the fact that the labeling of parents as egoistic or altruistic by Israeli counselors, was often not done in appreciation of the efforts put into raising a disabled child, but rather from the child and his sibling's point of view. Looking at the situation from this perspective, made insisting on the continuation of a "problematic" pregnancy seem like an egoistic act of a parent who resists abortion, while not considering the best interests of his existing and future children. As many Israeli counselors believe life can be "wrongful"(see Chapter 7) it is possible for them to justify their claim about the egoism of such parents by appealing to the rights of the future child, a right not to be born and treated as a subject, but rather the right to be aborted.

German and Israeli counselors were also strongly divided in their opinions concerning the duty of parents toward the siblings of children with disabilities (see Table 4).

While the majority of German counselors did not think it unfair to siblings to knowingly have a child with a disability (74.8 and 56.1%), only a minority among their Israeli counterparts felt the same (28.6 and 20.6%). The percentages of counselors agreeing with this claim were 7.1 times higher in Israel than in Germany in 1994–96 (61.9 compared to 8.7%) and 5 times higher in Israel than in Germany in 2000–01 (44.5 compared to 8.8%). Like in the case of the individual-society conflict, the "educational task" of the disabled was disregarded in Israel, but more appreciated in Germany, where a benefit was attributed to being a sibling of a disabled child.

TABLE 4. It is Not Fair to a Family's other Children Knowingly to Have a Child with a Disability

	1994–96				2000–01			
	Strongly Agree + Agree	Neither Agree nor Disagree	Disagree + Strongly Disagree		Strongly Agree + Agree	Neither Agree nor Disagree	Disagree + Strongly Disagree	
Germany	8.7%	16.4%	74.8%	100%	8.8%	35.1%	56.1%	100%
Israel	61.9%	9.5%	28.6%	100%	44.5%	34.5%	20.6%	100%

Finally, when asked to remark about the statement: I feel dissatisfied with counseling if patients decide to carry a seriously affected fetus to term, far more Israeli than Germans counselors agreed with this statement, revealing their more selective attitude.

Table 5 demonstrates that despite the non-directive ethos of genetic counseling, Israeli counselors feel much more dissatisfied with their counseling than their German colleagues, if the parents decide to give birth to a seriously affected fetus. Percentages of agreement with such a feeling were 8.7 and 8.5 times higher in Israel than in Germany: 42.8 compared to 4.9% in 1994–95 and 30 compared to 3.5% in 2000–01.

Up to now I have explained the differences between the studied groups of genetic counselors by referring to social ideals concerning family obligations and generational relationships. Another line of explanation for such differences has to do with diverse understandings of the parents' decisions to be either a mark of a backward culture, or rather a sign of a firm and appreciated opposition to the dangers science poses to a humanistic ethos. While a cultural meta-theme prevalent in Germany, understands science and progress and especially modern human genetics, to be in conflict with a humanistic ethos, as was clearly argued both by the German president Johannes Rau and by Jürgen Habermas, among many others, in the Israeli society, voices that reject modern science on the basis of it endangering humanity are hardly heard. As was mentioned previously in this work (see Chapter 3), in opposition to Germany where science is feared and criticism of scientific progress is very widespread, in Israel, science is still very much trusted and appreciated, as it is commonly believed that nature and fortune should be altered by it. Therefore, people who do not use the possibilities opened up by scientific innovations are not understood to be moral heroes, but rather, in the words of the counselors, "backward primitives", who put their trust in faith and God and not in knowledge.

Talking about their own personal and professional ethos, the Israeli counselors I interviewed repeatedly mentioned their strong trust in science and its ability to "correct" nature's problems and defects, contrasting it with a fatalistic attitude towards life. For example, a female Ph.D. in genetic counseling, working in one of Israel's largest hospitals said:

TABLE 5. Dissatisfaction with Parents who Decide to Carry to Term a Seriously Affected Fetus

	1994–96				2000–01			
	Strongly Agree + Agree	Neither Agree nor Disagree	Disagree + Strongly Disagree		Strongly Agree + Agree	Neither Agree nor Disagree	Disagree + Strongly Disagree	
Germany	4.9%	23.1%	72%	100%	3.5%	15.8%	80.7%	100%
Israel	42.8%	9.5%	47.7%	100%	30%	30%	40%	100%

"I don't accept destiny as it is. I am not a fatalist. Rather, as a scientist, I believe in managing our lives and controlling them..."

Her colleague, a female MD and a counselor who rightly felt herself to be holding a minority position said:

"Most people think they are civilized if they don't let uncertainty take control. Fatalism is equated with being backward, uncivilized and primitive. We are slaves to progress and I don't accept it, but this is just how it is. I understand there is no way to withdraw from where our civilization is now, but I don't think it should be sanctified".

To conclude, parents who choose to give birth to a disabled child are less appreciated among the Israeli counselors than among their German counterparts, who despite their profession's inclination for prenatal selection, value the parents' decision to resist it.

CONCLUSION

This chapter looked into potential conflicts between individuals and society and individuals and their families regarding reproductive genetics. Its findings demonstrate that whereas most Israeli counselors believe that the good society should be composed of a "healthy" population, many of their German counterparts believe the opposite to be true, in harmony with the dominant ideas of German public discourse. Thus, the conflict between individuals and society concerning this matter was formulated in a contradictory fashion. German counselors imagined a conflict between individuals' wish to avoid having abnormal children and society's need for genetic diversity and tolerance, as exemplified by the words of a German female counselor and gynecologist, working in the private sector:

"The conflict between the individual and society which is embedded in PND is that it aims at eliminating all diseases. From the point of view of the family this may be just, because it is very hard to have a child with a disability. Yet, society without the disabled is more limited because we must accept the disabled as part of our society, which makes life more complete".

In contrast, Israeli counselors imagined a conflict to result from certain parents' insistence on carrying problematic pregnancies to term, at the expense of burdening society. Thus, despite the Israeli counselors' surface liberal ideology, which glorified parents' autonomy, many of them viewed the decision to knowingly bring into the world children with genetic "defects" as socially irresponsible and understood the "good society" to be one free of the genetically abnormal.

This difference is further complicated by the two society's understandings of the potential conflict between the future child and her family, embodied in reproductive genetics. In accordance with the German common comprehension of society's interests, the repeating themes in the German discourse concerning parent's choice to accept children with genetic "defects", are that those parents are moral and psychological heroes, since they undertake their parental duties, which are to accept their children as gifts, they treat their children as ends and not as a means to satisfy their own needs or desires, they pay respect to natural processes and do not try to scientifically intervene with nature at the expense of humanity, and they do not view any form of life as wrongful. By contrast, in Israel, parents' choice to carry problematic pregnancies to term was mainly understood by genetic counselors not only as a social wrong, but also as a sign of egoism, fatalism and a backward culture. This has to do with the reality that the designing of future generations is not presented as a necessary moral wrong, or as a threat to the rights of the future child. Rather, it is seen as a sign of responsible parenthood. Additionally, as the concept of wrongful life is widely accepted, and family members are not viewed as individuals holding separate rights, but as parts of a group with similar interests, expectations regarding their mutual obligation are different in Israel than in Germany.

However, German counselors are generally more critical of their society's public discourse concerning the conflict between individuals and society and individuals and their mothers and families, embodied in the practice of reproductive genetics. This was clearly shown in this chapter by some of the German counselors' feelings that an open moral discussion concerning reproductive genetics is blocked in contemporary Germany, due to its Nazi past.

EPILOGUE

BIOPOLITICS AT THE BEGINNING OF LIFE

The Foucauldian model of biopower emphasizes the context of modernity in which the human body has become a target and means of political strategies named biopolitics. According to Foucault, medicine (like sexuality) serves as a hinge between the level of the private body of individuals and the body of the population or the species, both of which are policed, supervised and examined for their condition and its improvement and protection. Thus, medicine polices private bodies at the same time it regulates populations (Prainsack, 2004). This Foucauldian model has been extended and updated by different writers. For example, Nikolas Rose (2001) contends that the by now classical Foucauldian terms, are no longer relevant to "advanced liberal societies" for different reasons. Among them is the claim that biopolitics can no longer be framed in terms of the "fitness of the nation" because the care for health has been individualized. A good example is the case of reproductive genetics which is no longer coerced by the state. Rather, individual patients are expected to make "autonomous decisions" as part of their self-governance according to medical norms. Likewise Rose contends that nowadays biopolitics is driven by economic forces, a process which leads to the breaking down of traditional classifications between cure and enhancement (a good example is plastic surgery) and between commodities and that which is human (an apposite example is selling and buying human organs). Thus, contemporary biopolitics according to Rose disputes the value to be accorded to "life itself", the "quality of life", the "right to life" and so on. Likewise Giorgio Agamben (1998) claims that the fundamental biopolitical structure of modernity is the sovereign's decision on the value or **non value** of life as such in different social spaces. According to him what is unique about modern politics is not merely its inclusion of the human body in the polis, but rather the fact that the hidden foundation upon which the entire political system rests is "at once excluding bare life from and capturing it within the political order" (p. 9). Whereas Foucault speaks about the management of life, Agamben speaks of the "state of exception" in which death and life interface each other. Thus, it is exactly the valorization and politicization of life that is characteristic of modernity, which inevitably implies a locus where life ceases to be politically relevant and becomes "bare life", or the life of what he terms "*homo sacer*", the person whose life can be eliminated without punishment or killed but not murdered or sacrificed. His paradigmatic example for such life, or for such a state of exception, is the life of those imprisoned in concentration camps.

In the realm of medicine, the ability to manipulate the material of life through contemporary technologies has destabilized the very notion of life itself (Rabinow, 1992, 1996). Equally, the dramatic rise in the medical ability to sustain the life of the genetically abnormal, combined with the even newer ability to detect these "lives" in-uterus and hence, to prevent them, has created a social need for new definitions of "life". However, this need has not been widely met by a comparative point of view that is sensitive to cultural differences and their interaction with contemporary technologies.[1]

Building on former theories concerning biopolitics and understanding reproductive genetics as being especially fit to be understood as a tool for deciding about the value or non value of prospected life, or as a locus for dispute about the value of "life itself" (see also Sawicki, 1999), I contend that the theories of Foucault, Rose and Agamben, with all their differences and similarities, and despite their acknowledgment that "life" is not merely a biological/scientific concept, are blind to differences between cultures. They do not draw our attention to the fact that biopower, namely the power of modern societies over life (and following Agamben, over death), manifests itself in different ways in various modern nations. Thus, claiming that former theories of biopower/biopolitics are limited in allowing a culturally-specific understanding of the workings of this power, in this book I aimed to provide a nuanced and "culturally sensitive" application of the concept of biopolitics/biopower. Thus I pointed to its cultural designation and claimed that biopower is more accurately described as dependant upon vernacular interpretations of the concept of "life" itself, which dialectically constructs biopower and is regulated by it.

Thus, the objective of this study has been to pursue a focused analysis of genetic reproductive technologies in Israel and Germany and their relation to local understandings of the notion of "life", with a special emphasis on the hegemonic conceptualization and metaphorization of the beginning of "life". The central questions of this study concerning the communities in question, had thus become; When does life begin? How wide is the border between pre- and postnatal life? When is life understood to be (un)worthy of living and thus, what is the purpose of life and how legitimate is the technological manipulation of life itself?

Studying these questions I have claimed that the concept of "life", and consequently its entitlement (or non entitlement) for care and protection, is organized according to hegemonic cultural perceptions concerning the human natural purpose, the moral standing of fetuses, society's alleged interests, the hopes and dangers embodied in science and technology and the ideal relationship between human generations.

In the following, I attempt to summarize my findings concerning the unique biopolitics of the beginning of life characteristic of the present societies in Israel and Germany. Supporting the different uses of reproductive genetic technology in Germany and Israel, this study described a matrix of social, cultural, legal, historical,

[1] More comparative work has been done on end of life decisions. See for example Lock's discussion of the concept of brain death in Japan and the US (Lock, 2002). For a comparative discussion of fetus' personhood see Gross, 1999 and Conklin and Morgan (1996). For a comparative disussion of the quality versus the sanctity of life in the context of euthanasia, see Cohen-Almagor and Shmueli (2000).

and political differences, which conditioned the formation of the two different medical-genetic fields studied. The following section points to these differences, without placing them in an hierarchical or temporal order, since this study has consistently claimed and demonstrated that cultural and theological repertoires feed the medical, legal and political systems and vice versa, in an inseparable manner. This also explains why the themes represented in the following section partly intertwine and overlap.

LIFE AND ITS DISTINCT CULTURAL MEANINGS

The Major Differences between contemporary Israeli and German societies, relevant to the practice of reproductive genetics and to their biopolitics of the beginning of life, can be summarized along the lines of six different themes: the importance of fertility, the moral standing of the fetus, images of the ideal relationship between generations, the notion of wrongful life, eugenic history and its lingering effects, as well as attitudes towards science and technology and their ability to manipulate "life itself".

Firstly, In Israel fertility is greatly valued and Israel's total fertility rates are high compared to all other "advanced liberal societies" (Berkovitch, 1999; Fogiel-Bijaoui, 1999). Common explanations for this include identification with the collective goal of fighting the "demographic threat" and Arab enemies (Yuval-Davis, 1987), the threat of losing a child in war or in a terrorist attack, Jewish genocide during the holocaust (and the consequent wish to replace the dead) and Jewish tradition (Kahn, 2000), in which parenthood is an important moral command and infertility is treated as a severe disability (Marx, 2002). Consequently, NRTs are very popular and strongly supported by the state (Haelyon, 2004; Kahn, 2000). Additionally, and much like in the rest of the advanced world, pregnancy is intensely medicalized in Israel (Remennick, 2000; Sered, 2000).

In contrast, Germany's total fertility rates are among the lowest in the world and German society is often referred to by its own citizens as hostile towards children (Keller, Zach and Abels, 2005). Remaining childless (or rather choosing to stay "childfree") is considered a legitimate personal choice in Germany, as there exists no "fertility craze". Furthermore, while the German decline in fertility is widely understood to threaten the economy, it is not perceived as threatening the existence of the German state in its present political form (Moeller, 1993). Thus, although pregnancy is obviously also medicalized in Germany, reproductive medicine is not very "fashionable". In fact, a lot of NRTs are prohibited and this does not raise social protest.

Secondly, both the legal system in Israel and Jewish doctrine understand life to begin after birth, since fetuses are perceived as parts of their mothers with no autonomous rights (Amir, forthcoming; Gross, 1999; Jakobovits, 1967; Morag-Levine, 1994; Steinberg, 1991; Yishai, 1993). On the other hand, in Germany embryos (as well as pre-embryos) are perceived as individuals holding autonomous rights (German National Ethics Council, 2003).

The roots of this attitude can be traced to the Christian (mostly Catholic) tradition, which influences the German hegemonic perception of the fetus as a "life", as well as to the history of the "Third Reich". According to Wuerth (1997) the Nazi past is responsible for the moral position of the unified German state, which protects **all** life, as a reaction to the past and as a hallmark of the new "morally rehabilitated" Germany. Thus, the German Embryo Protection Act (1990) (which has no parallel in Israel), limits the use of NRTs in Germany, as it highly protects the very early stages of pregnancy and confers rights on very young fetuses (even those formed in the laboratory, outside a woman's body). As a result of this stance, a number of technologies are prohibited, namely: egg and embryo donation; attempting to fertilize more eggs from a woman than can be transferred back into her within one IVF treatment cycle, (in order to protect against intentional embryo wastage); surrogate motherhood; pre-implementation diagnosis; stem cell research;[2] sex selection (in the absence of sex-linked inheritable diseases); fertilization with sperm retrieved post-mortem and cloning. Except for cloning (Prainsack, 2006a), all these techniques are permitted in Israel. Thus, whereas in Germany stem cells and pre-embryos as well as more developed fetuses, are considered to hold the rights entitled to "life", no such rights are conferred to them under Israeli law (Strasberg-Cohen, 2005) or in the Jewish tradition.

Thirdly, Israel is considered a pro-natalist society in which public and private lives revolve around children (Portugese, 1998). However, just like embryos who are perceived as first and foremost parts of their mothers, children in Israel are perceived as parts of a group (family, nation), and therefore are expected to fulfill the group's wishes for "healthy" and fertile offspring. Emphasis is placed on the rights of parents, family and present society, over and above the rights of the future child or of future generations (Landau, 1996). Furthermore, children are expected to fulfill the needs and even desires of their family.

In contrast, in Germany children are considered as "gifts" (Rau, 2001), which must be accepted by their receivers (family or society), in order to maintain and display the receiver's morality. Emphasis is placed on the rights of future generations or individuals and not on the rights of mothers, parents and siblings. Following Kantian philosophy as well as the contemporary writings of Jürgen Habermas, children must be seen as ends and never as means and genetic enhancement of children is strongly opposed on the basis that it threatens the natural moral relationship between generations (Habermas, 2003; Krones and Richter, 2004; Prainsack and Spector, 2006).

Fourthly, according to the Israeli legal system, some kinds of life are considered "wrongful", (as the Israeli Supreme Court accepted a wrongful life suit). Accordingly, Israeli law explicitly allows eugenic abortions and in practice, it is very easy to abort for eugenic reasons in Israel, even after viability. In contrast, German courts did not accept "wrongful life" suits, which are understood to threaten a

[2] Since 2002 research on stem cells that have been imported to Germany prior to a certain deadline is allowed (Lanzerath, 2004).

humanistic ethos and to devalue the life of the disabled (Harrer, 1994). Equally, German law does not explicitly allow eugenic abortions. (Yet, such abortions are carried out in Germany as abortions on demand are non-punishable in the early months of pregnancy. After viability, German women can use the "mother's health clause" as an excuse for performing late selective abortions. However, in practice, late selective abortions are difficult to obtain in Germany).

In addition, even disability activists in Israel support eugenic ideas, namely using reproductive genetics to prevent life with disability. Raz (2004) explains this to result from them not fearing that this position will threaten the care for the living disabled, due to the strong differentiation they make between pre- and postnatal life. This attitude stands in sharp contrast to the prevalent position of disability organizations in Germany, which claim that selective abortions devalue the life of the living disabled. Likewise, repeating themes in the German public as well as disability discourse use positive stereotyping to describe the disabled, who are believed to live a "richer" life and to serve society, as they make it more humane. This theme also echoes Christian understandings of the meaning of suffering, both for the individual person and for society on the whole (Steinberg, 1999). In contrast, according to Marx (2002), attitudes to disability found in Jewish thinking are often disturbing to contemporary sensibilities, as Jewish culture is to a large extent a competence-oriented culture of "learners", which excludes and stigmatizes the disabled. Concerning contemporary Israeli society, Weiss, in her thesis about the Israeli collective body ideal, which she terms the "chosen body", contends that it worships healthy, fit, competent and whole bodies. Accordingly, the disabled are negatively stereotyped in the Israeli culture (Remennick, 2006; Weiss, 2002). Likewise, suffering is not glorified or endowed with personal or collective meanings and the life of the severely handicapped is believed to be wrongful.

Fifthly, Jewish tradition supports the prevention of life with disability, especially prior to conception (Jakobovits, 1967). No such "soft" eugenic tendencies can be found in Christian tradition, either in the past or in the present (Bassett, 1995). Based on Jewish tradition concerning the health of offspring, the nowadays ultra-orthodox Jewish community is utilizing contemporary genetic knowledge in a unique fashion. It runs a special program called "*Dor Yeshorim*", for carrier screening tests prior to prearranged marriages, (the way to marry in this community) and thus prevents the conception of children with diverse autosomal recessive disorders (especially Tay-Sachs; see Prainsack and Siegal, 2006). Likewise, contemporary Israeli medical genetics is not haunted by the shadow of Nazi eugenics, but rather supported by the relatively "soft" Jewish-Zionist eugenic history (Falk, 2002; Hirsch, forthcoming; Stoler-Liss, 1998) and by a general favorable attitude toward science and medicine found in the Israeli-Jewish culture. Thus, whereas the biologically murderous history of the Nazi-period is very rarely related to genetics in Israel (Kirsh, 2003), in Germany guilt feelings resulting from it have led to a highly emotional moral discussion. Herzog (1998) argues that the thematic focus of the present German political identity was and still is one of dissociation from the Nazi past, as in contemporary public discourse in Germany, leftists and conservatives alike, deploy holocaust images and memories as a sledgehammer technique in their political battles. In this context, regulation of the private body in Germany today is highly affected by the Nazi regimes' handling of "unworthy lives", either for racial or for genetic reasons. In view of that, Linke (1999) explains, for example, the German Green movement's concern for the disabled and the mentally ill, to be based on an attitude which runs counter to the horrors of Nazism, where "mercy killing" was the fate of the disabled and where Nazi physicians were the founders of the logic and technique behind the final solution. Consequently, it is the aim of German bioethics today to fight for the rights of those populations that had previously been labeled "inferior". As a result, the German "good society" of our days is ideally seen as genetically varied, quite the opposite from the Israeli "good society", which is ideally seen as genetically "healthy".

Sixthly, being the prototype of a risk society (Beck, 1992; Cohen, 1999; Giddens, 1991), Germany is characterized by a strong distrust of science and technology. Apocalyptic fears formerly channeled at nuclear energy in Germany are now directed towards genetic research, as contemporary German survival fears are centered on probable distant risks. In contrast, Israeli society is generally trustful of science and technology, which is perceived to guarantee its existence in the Middle-East (Prainsack and Firestine, 2006). Thus, individual and collective concerns about survival are channeled at the prevention of concrete and tangible risks (stemming from war or terrorism) and not at risks stemming from the technological era. Moreover, in contrast to the German hegemonic moral discourse, the dominant Israeli-Jewish culture does not perceive the technological manipulation of "life" itself as either "playing God" or as threatening to human dignity or rights. Alternatively, the prevention of life with disability is not seen as endangering a humanistic ethos but as preventing suffering and improving on God's creation, as advanced medical technologies are understood to serve the common good and not to pose ecological or moral/theological risks or sins.

To conclude, no groups opposed to preventing the birth of genetically abnormal children exist in Israel, as criticism is muted. Even the ultra-orthodox Jews (a segregated minority) oppose abortions (unless the embryo or fetus poses a danger to the health or life of the potential mother) but support the preconception prevention of life with disability. Similarly, disability activists in Israel do not reject reproductive genetics. On the other hand, in Germany mixed opposition groups and political coalitions all contest reproductive genetics, such as members of different parties (Greens, SPD and CDU), disability activists, feminists, environmentalists and religious thinkers. Accordingly, in Israel reproductive genetics is pushed forward by the state, as well as by the medical establishment, unlike in Germany where the state and even more so the medical establishment, limit the possible uses of reproductive genetics. Thus, the current options opened up by reproductive genetics, namely of closing the social doors to the genetically abnormal, or allowing future parents to avoid having disabled children, are much more extensively used in Israel than in Germany.

ZOĒ AND *BIOS* BIOPOLITICS AT THE BEGINNING OF LIFE

Present technological possibilities for manipulating the very biological processes called "life", imply a biopolitics that contests how, when and for what purposes such technologies should be used (Dean, 2004). Up to now I have traced the cultural, social and political powers involved in excluding and including certain forms and stages of "life" in Germany and Israel. Accordingly, I demonstrated that whereas the selection among embryos in both Israel and Germany is justified along the logic of a spectrum leading from worthy and protected life to unworthy and hence not protected life, or even a life which should be destroyed, the placing of different lives upon this spectrum is poles apart in Israel and Germany.

The last section of this book is dedicated to questioning how the fate of fetuses who exist on the border of life at our present technological age, may be comprehended in light of Agamben's political philosophy. Thus, my question is how different, so called advanced liberal societies, such as Israel and Germany, incorporate *zoē* (the simple fact of living, or biological life *per se*, common to all living beings, animals, men or gods), and *bios*, (political life, namely the form or way of living proper to an individual or a group), within their biopolitics, or how their balancing of *zoē* and *bios* affect their differential practices of reproductive genetics.[3]

Agamben argues that the *Homo Sacer*, the "sacred" human being who is in a way reduced to his or her "bare life", does not exist only in concentration camps or in boats filled with refugees, but is also the outcome of contemporary medical technology. He explains that the modern practice of euthanasia demands a separation of *zoē* and *bios* in another person, or isolation of some sort of "bare life" that may be killed without having to fear persecution. Thus, the patient who lives on life support and wavers between life and death, similarly to the Muselmann in the concentration camp, is an extreme embodiment of bare or naked life, the life of the *homo sacer*, whose biological life is separated from its *bios*, or biography (Agamben, 1998; Wynn, 2002). Such a separation is paralleled in the case of selective abortions, which likewise demand a distinction between qualified life and "bare life" in a fetus which does not yet have a biography but is in the process of obtaining one.

Agamben's philosophical project is to re-think the risky separation between *zoē* and *bios* underlying political systems in the West (Wynn, 2002). Thus he wishes for political life to aim at not separating life from what he calls "form of life", namely a life from which bare life cannot be separated, or again, a life in which *zoē* is inseparable from its *bios* (Agamben, 1993). Returning to reproductive genetics, Dean (2004) writes: "Sovereignty – the power of killing – is today practiced in the biomedical domain by health professionals and administrators, by relatives and carers, and by prospective parents and mothers, all under the watchful guardianship of institutional ethical committees, legal regulation and therapeutic expertise" (p. 19). My following question is, how can we make sense of this killing

[3] In the philosophical tradition, it was Hannah Arendt (1998) who restored the Greek/Aristotelian terms *zoe* and *bios* in her discussion of life.

in light of Agamben's philosophy? I argue that while it is perfectly clear what it means to fight for life in order to become a "form of life" once again (in which *zoē* and *bios* are inseparable) in refugee boats or concentration camps, when dealing with end or beginning of life decisions, what Agamben implies is no longer as obvious. I thus wish to ask whether Agamben's view leads to a radical condemnation of contemporary reproductive genetic practices, or not. Arguing along the lines of Dean (2004) and Wynn (2002), I suggest that the idea of mapping the zones of *bios* and *zoē* and their indistinctness in the context of reproductive genetics, may facilitate fertile intellectual contestations and a complex analysis, rather than a dogmatic rejection of this practice.

Throughout this book I have demonstrated that the hegemonic understandings of the purpose of life, the beginning of life, the quality versus sanctity or value of life, the legitimate manipulations of life, as well as of the relationship between human generations embodied in the creation of life, profoundly vary in contemporary German and Israeli societies. Thus, I have shown that in Israel a fetus is considered culturally (and hence medically) viable only if it is potentially "healthy", "fertile" and "normal" and not otherwise, and thus its viability may be denied. On the other hand, in Germany viability is understood to begin earlier and to include both very early fetuses and abnormal fetuses throughout the pregnancy. In other words, the current German biopolitics of the beginning of life is devoted to the simple act of living itself. Hence, it may be understood as a case in which the *bios* has swallowed up its own *zoē*, to become indistinguishable from it and thus, every stage or form of "life" is protected or made to survive and every life or promise of a life becomes *bios*: life proper to members of society. On the other hand, the biopolitics of the beginning of life characteristic of the Israeli society can be understood as a case in which the *zoē*, or the simple act of living, is constantly being separated from its *bios*, in order to allow for its elimination. Hence certain stages or forms of "life" may be unprotected. Consequently, the Israeli hegemonic moral order concerning qualified life may be described as a "*bios* biopolitics" in which "life worthy of living" must be characterized by more than the simple act of living. On the other hand, the current German hegemonic moral order may be described as a "*zoē* biopolitics", in which every type of "life" (or pre-life) is considered politically qualified. However, in the realm of medical technology, the radical non separation of *zoē* and *bios*, or the insistence on mere survival, may be morally problematic, as it is medical technology itself which is often guilty of separating *zoē* and *bios* in a person. Thus, in some cases, such as those of neonates at the edge of viability (Wynn, 2002), or of severely handicapped fetuses who will be heavily dependant on medical technology in order to survive, letting die, rather than insisting on survival, may be a deep recognition of "life" as a form of life.

To conclude, the new terms "*zoē* biopolitics" and "*bios* biopolitics", which I offer, develop the concept of biopolitics by stressing that it is not one and the same in all modern societies but rather, cultural-specific forms of biopolitics exist within contemporary modern technological societies. In line with the general thesis of Agamben, these concepts do not point to the possibility of freeing the human body from its subjection to politics, as I too understand it to be no longer possible to simply give back the body of the West (or of mothers and fetuses) to its "natural" life outside politics. Accordingly, I too understand the German and Israeli biopolitics' handlings of life as two forms of modern governance, which is centered on the unity of life and politics, without which it is incomprehensible, as both societies make use of their offspring in order to justify and glorify the existence of their community. However, reading the story of reproductive genetics in both countries through the theoretical prism of biopolitics offered by Foucault and elaborated by Rose and Agamben, does not mean that I think the differences between the diverse biopolitics of the beginning of life described in this book are meaningless. Rather, by describing the different interplay between *zoē* and *bios*' incorporation within different societies' biopolitics, I resist Agamben's conclusion, which does not differentiate between diverse forms of what he terms totalitarianism. On the contrary, I think that in order to better understand the complexity of modern genetics and its handling of "life" it is important to understand that although any redefinition of the bio-political borders between life and death indicates that the exercise of sovereign power passes through them, important variations are present within modern societies' biopolitical subjection of bare life to the polis, and its protection or exclusion. My aim in this book has not been to provide a moral definition of life or to place different definitions of life on a moral scale. Rather, I tried to point at "life's" cultural designation, or at it being a philosophical, political, and theological concept, rather than a medical/scientific one and thus, to reengage sociology with fundamental questions regarding the social meaning of life and death (Shilling and Mellor, 2001). I believe that doing so in regard to reproductive genetics is important for understanding the concrete effects it has on the management of genetically abnormal life in Israel and Germany, as the interplay between local meanings of "life" and the common ground of a modern rational scientific discourse is responsible for the different ways in which genetic technology is applied in different societies. Thus, studying such differences is important for understanding the complex liberties and constraints, blessings and burdens, embodied in reproductive genetics for fetuses and their mothers and families, for the handicapped and for society as a whole.

Building on the conceptualization of "life" as manifested in the discourse of reproductive genetics in Israel and Germany, let me now briefly move to a broader discussion of the handling of "life" in our modern technological era. Thinking of different cultural moral orders concerning "life" in terms of the relationship between *bios* and *zoē*, may serve for the study of other situations in which contemporary societies have to decide about the borders between life and death and about future technological manipulations of life and their effects on the nature of human community and on what it means to be human. Relevant technologies, not necessarily genetic ones, would have to do with the medical capacity to sustain life, transplant organs, enhance life or clone it. All such techniques call for a rethinking of the concept of life itself. Thinking of such issues in the light of the analysis offered in this book may allow for a less culturally-bound understanding of them.

APPENDICES

APPENDIX 1: THE GENETIC CONDITIONS

The genetic conditions are

1. Mother's life in danger **2**. Pregnancy result of rape **3**. Hurler Syndrome **4.** Trisomy 13 **5**. Cleft lip and palate (girl) **6**. Severe, open spina bifida **7**. Cystic Fibrosis **8**. Anenchephaly **9**. Sickle cell anemia **10**. Familial hypercholesterolemia, homozygous **11**. Predisposition to schizophrenia or bipolar disorder **12**. Huntington disease **13**. Predisposition to Alzheimer **14**. Predisposition to alcoholism **15**. 45,X **16**. Trisomy 21 **17**. XXY **18**. Cleft lip and palate (boy) **19.** Neurofibro-Matosis **20**. Severe obesity (in absence of a known genetic syndrome) **21**. Achondroplastic dwarfism **22.** Child is not the sex desired **23.** HIV infection in the fetus **24**. Toxoplasmosis in 1st trimester **25**. Rubella in the fetus (1st trimester) **26**. PKU.

APPENDIX 2

How Would You Counsel? For and Against Abortion

	1993–95			2000–01		
	Germany Mean (s.d.)	Israel Mean (s.d.)	t-Test	Germany Mean (s.d.)	Israel Mean (s.d.)	t-Test
The condition						
1. Mother's life in danger	4.24 (0.92)	4.29 (0.90)	−.233	4.37 (0.88)	4.55 (0.87)	.361
2. Pregnancy result of rape	3.17 (0.70)	3.76 (0.70)	−3.638**	2.91 (0.49)	3.31 (0.66)	−2.892**
3. Hurler Syndrome	3.55 (0.73)	3.86 (0.65)	−2.010	3.38 (0.64)	4.15 (0.77)	−4.810**
4. Trisomy 13	3.96 (0.88)	3.81 (0.60)	1.000	3.64 (0.81)	4.37 (0.67)	−4.234**
5. Cleft lip and palate (girl)	1.90 (0.78)	2.90 (0.30)	−11.53**	1.90 (0.85)	2.76 (0.69)	−5.07**
6. Severe, open spina bifida	3.64 (0.78)	3.95 (0.67)	−2.018	3.47 (0.65)	4.20 (0.71)	−4.836**

(Continued)

How Would You Counsel? For and Against Abortion (*Continued*)

	1993–95			2000–01		
	Germany Mean (s.d.)	Israel Mean (s.d.)	t-Test	Germany Mean (s.d.)	Israel Mean (s.d.)	t-Test
7. Cystic Fibrosis	3.16	3.52	– 2.610*	3.00	3.63	−3.452**
(0.60)	(0.60)		(0.65)	(0.89)		
8. Anenchephaly	4.31	4.05	1.352	3.97	4.53	−3.241**
(0.86)	(0.80)		(1.01)	(0.63)		
9. Sickle cell anemia	2.97	3.43	−3.061**	2.86	3.18	−1.730
	(0.65)	(0.60)		(0.81)	(0.77)	
10. Familial hypercholesterolemia, homozygous	2.64	3.19	−3.062**	2.59	2.97	−1.976*
	(0.79)	(0.60)		(0.88)	(0.81)	
11. Predisposition to schizophrenia or bipolar disorder	2.41	3.00	−6.359**	2.40	2.71	−1.437
	(0.82)	(0.32)		(1.01)	(0.85)	
12. Huntington disease	3.01	3.71	−4.290**	2.84	3.43	−3.468**
	(0.72)	(0.64)		(0.83)	(0.57)	
13. Predisposition to Alzheimer	2.46	3.10	−4.218**	2.32	2.73	−1.906
	(0.87)	(0.62)		(1.00)	(0.91)	
14. Predisposition to alcoholism	2.32	3.00	−6.785**	2.24	2.60	−1.823
	(0.96)	(0.32)		(0.86)	(0.89)	
15. 45,X	2.20	3.30	−9.206**	2.28	3.03	−4.785**
	(0.76)	(0.47)		(0.77)	(0.67)	
16. Trisomy 21	3.15	3.86	−4.770**	3.05	3.57	−2.990**
	(0.64)	(0.65)		(0.54)	(0.86)	
17. XXY	2.30	3.05	−6.192**	2.24	3.03	−5.283**
	(0.73)	(0.50)		(0.76)	(0.61)	
18. Cleft lip and palate (boy)	1.85	3.00	−10.06**	2.03	2.57	−2.908**
	(0.80)	(0.45)		(1.03)	(0.68)	
19. Neurofibro-Matosis	2.71	3.19	−4.561**	2.64	3.00	−2.606*
	(0.76)	(0.40)		(0.77)	(0.53)	
20. Severe obesity (in absence of a known genetic syndrome)	2.42	2.43	−0.63	2.18	2.67	−2.455*
	(0.88)	(0.87)		(0.87)	(0.92)	
21. Achondroplastic dwarfism	2.86	3.62	−4.854**	2.77	3.53	−4.292**
	(0.68)	(0.67)		(0.63)	(0.86)	
22. Child is not the sex desired	1.86	2.24	−1.038	1.60	1.73	−0.610
	(1.63)	(1.22)		(1.05)	(0.78)	
23. HIV infection in the fetus	3.24	4.05	−3.814**	2.91	3.64	−3.846**
	(0.93)	(0.80)		(0.82)	(0.83)	

24.	Toxoplasmosis in 1st trimester	2.95 (0.89)	3.86 (0.79)	−4.474**	2.85 (1.01)	3.90 (0.76)	−4.962**
25.	Rubella in the fetus (1st trimester)	3.48 (0.83)	4.10 (0.77)	−3.245**	3.34 (0.84)	3.93 (0.70)	−3.265**
26.	PKU	2.38 (0.80)	2.95 (0.97)	−3.044**	2.20 (0.83)	3.00 (1.04)	−3.862**

Question 34: How would you counsel, in your professional capacity, about termination of pregnancy for each of the following conditions? Assume that a reliable test has been developed. *Answers*: Not tell them this particular test result + urge parents to carry to term (1), Emphasize positive aspects so they will favor carrying to term without suggesting it directly (2), Try to be as unbiased as possible (3), emphasize negative aspects so they will favor termination without suggesting it directly (4), urge termination (5).

How Would You Act Personally? (During First Trimester)

	1993–95			2000–01		
	Germany Mean (s.d.)	Israel Mean (s.d.)	t-Test	Germany Mean (s.d.)	Israel Mean (s.d.)	t-Test
The condition						
1. Mother's life in danger	2.93 (0.27)	2.95 (0.22)	−0.258	2.92 (0.26)	3.00 (0.00)	−2.057*
2. Pregnancy result of rape	2.69 (0.47)	3.00 (0.00)	−8.301**	2.60 (0.49)	2.96 (0.19)	−4.585**
3. Hurler Syndrome	2.89 (0.31)	3.00 (0.00)	−4.607**	2.75 (0.43)	2.96 (0.19)	−3.059**
4. Trisomy 13	2.94 (0.23)	3.00 (0.00)	−3.248**	2.87 (0.34)	3.00 (0.00)	−2.810**
5. Cleft lip and palate (girl)	1.40 (0.58)	2.60 (0.50)	−8.605**	1.54 (0.61)	2.40 (0.69)	−5.711**
6. Severe, open spina bifida	2.88 (0.34)	3.00 (0.00)	−4.493**	2.75 (0.43)	3.00 (0.00)	−4.111**
7. Cystic Fibrosis	2.56 (0.54)	3.00 (0.00)	−10.65**	2.38 (0.63)	2.89 (0.40)	−4.443**
8. Anenchephaly	2.95 (0.28)	3.00 (0.00)	−2.16*	2.90 (0.29)	3.00 (0.00)	−2.326*
9. Sickle cell anemia	2.30 (0.60)	2.90 (0.22)	−9.46**	2.21 (0.70)	2.78 (0.50)	−4.258**
10. Familial hypercholesterol-emia (homozygous)	2.09 (0.67)	2.75 (0.44)	−4.235**	1.94 (0.66)	2.55 (0.63)	−4.071**

(Continued)

How Would You Act Personally? (During First Trimester)

		1993–95			2000–01		
		Germany Mean (s.d.)	Israel Mean (s.d.)	t-Test	Germany Mean (s.d.)	Israel Mean (s.d.)	t-Test
11.	Predisposition to schizophrenia or bipolar disorder	1.76 (0.61)	2.47 (0.68)	−5.019**	1.66 (0.68)	2.19 (0.80)	−3.085**
12.	Huntington disease	2.35 (0.63)	3.00 (0.00)	−13.44**	2.22 (0.69)	2.93 (0.26)	−6.714**
13.	Predisposition to Alzheimer disease	1.76 (0.58)	2.4 (0.75)	−3.621**	1.53 (0.66)	2.06 (0.80)	−3.236**
14.	Predisposition to alcoholism	1.60 (0.55)	2.35 (0.67)	−5.656**	1.5 (0.61)	1.92 (0.72)	−2.850**
15.	45,X	2.00 (0.57)	2.85 (0.36)	−6.696**	1.92 (0.48)	2.86 (0.44)	−8.698**
16.	Trisomy 21	2.70 (0.47)	3.00 (0.00)	−8.361**	2.60 (0.53)	3.00 (0.00)	−5.429**
17.	XXY	1.98 (0.57)	2.90 (0.30)	−7.249**	1.77 (0.57)	2.75 (0.51)	−7.724**
18.	Cleft lip and palate (boy)	1.45 (0.59)	2.60 (0.50)	−8.363**	1.44 (0.54)	2.50 (0.64)	−7.900**
19.	Neurofibro-matosis	2.32 (0.55)	2.61 (0.67)	−1.95	2.09 (0.66)	2.85 (0.36)	−6.668**
20.	Severe obesity, in absence of a known genetic syndrome	1.86 (0.71)	2.38 (0.80)	−3.081**	1.62 (0.59)	2.34 (0.77)	−4.362**
21.	Achondroplastic dwarfism	2.47 (0.60)	2.9 (0.31)	−5.161**	2.35 (0.65)	2.90 (0.40)	−4.771**
22.	Child is not the sex desired	1.08 (0.30)	1.65 (0.81)	−3.075**	1.14 (0.41)	1.30 (0.47)	−1.495
23.	HIV infection in the fetus	2.58 (0.63)	3.00 (0.00)	−8.704**	2.44 (0.63)	3.00 (0.00)	−6.435**
24.	Toxoplasmosis in the fetus in first trimester	2.48 (0.59)	2.95 (0.22)	−7.052**	2.38 (0.69)	2.93 (0.25)	−5.159**
25.	Rubella in the fetus in first trimester	2.82 (0.38)	3.00 (0.00)	−6.022**	2.37 (0.49)	3.00 (0.00)	−3.964**
26.	PKU	2.06 (0.61)	2.80 (0.41)	−5.245**	1.90 (0.62)	2.65 (0.56)	−5.243**

* *Sig* < 0.05 ** *Sig* < 0.01

Question 35: How would *you personally* respond if you yourself in your own life faced the possibility of having a child with the disorders listed below? (During the first 3 months of Pregnancy): I would have an abortion, I would not have an abortion, but it should be legal for others, I would not have an abortion and it should be illegal for others, 1–3 interval scale).

APPENDIX 3: INDEPENDENT VARIABLES

Independent variables	Description
Nationality	0= Israel 1= Germany
Sex	0= man 1=woman
Age	In Years
Religiosity	How important is religion in your life? 1=Not at all 2= Slightly 3=Moderately 4=very 5= Extremely
Place of Training	0=abroad 1=at home
Professional Experience	In years
Medical sub-specialty 1	0=other 1=pediatrician
Medical sub-specialty 2	0=other 1=obstetrician
Time of research	0= 1993–5 1= 2000–1

APPENDIX 4

The indexes were made up of the averages of answers to the following questions.

Descriptive Statistics of the indexes

PND Practices	Pro Selective Abortions	Pro Abortion Practices	Pro Eugenic Practices	Pro Disabled are Good for Society	Pro Parental Autonomy	Pro Rational Scientific Planning of Life
Mean	3.08	3.18	2.04	3.59	2.84	2.83
Standard deviation	0.27	0.92	0.67	0.77	0.54	0.74
α	0.88	0.46	0.73	0.49	0.64	0.62

In the cases of Pro Abortion and Pro Disabled are good for society, only two questions formed the indexes. Hence, the weak α.

Pro Selective Abortions

Question 34: How would you counsel, in your professional capacity, about termination of pregnancy for each of the following conditions? Assume that a reliable test has been developed. *Answers*: Not tell them this particular test result + urge parents to carry to term (1), Emphasize positive aspects so they will favor carrying to term without suggesting it directly (2), Try to be as unbiased as possible (3), emphasize negative aspects so they will favor termination without suggesting it directly (4), urge termination (5).

(In order for the answers to indicate pro selective abortions attitudes the numbers of optional answers were reversed 1 = 5, 2 = 4 3 = 3 4 = 2 5 = 1).

Pro Abortion Practices

1. A woman's decisions about abortion should be her own, without intervention by anyone 37 [12])
   Answers: Strongly Agree (1), Agree (2), neither Agree nor Disagree (3), Disagree (4), Strongly Disagree (5).
2. A woman should have an unqualified right to abortion before viability (in the original question 37 [30])
   Answers: Strongly Agree (1), Agree (2), neither Agree nor Disagree (3), Disagree (4), Strongly Disagree (5).

(In order for the answers to indicate pro abortion attitudes the numbers of optional answers were reversed 1 = 5, 2 = 4 3 = 3 4 = 2 5 = 1)

Pro Eugenic Practices

A single blind woman on public welfare has a dominant genetic disorder that leads to legal blindness. She has three sons. Each has a different father. All three fathers are absent. All three sons are 90% blind in spite of early attempts at treatment. The mother tells the geneticist: "Now I think I'll have a little girl." There is a 50% risk that the child will be blind. The geneticist thinks privately, "Don't you think you have brought enough suffering into the world?"

For each statement below, indicate the extent to which you agree or disagree. (In the original question 30 [2,3,4,5])

The geneticist should convey this sentiment to the woman, The geneticist should try to persuade her not to have another child, The geneticist should suggest sterilization, There should be a law requiring her to be sterilized.

Answers: Strongly Agree (1), Agree (2), Neither Agree nor Disagree (3), Disagree (4), Strongly Disagree (5).

An important goal of genetic counseling is to reduce the number of deleterious genes in the population. (37[14])

Answers: Strongly Agree (1), Agree (2), Neither Agree nor Disagree (3), Disagree (4), Strongly Disagree (5).

(In order for the answers to indicate pro eugenic attitudes the numbers of optional answers were reversed 1 = 5, 2 = 4 3 = 3 4 = 2 5 = 1)

Pro Disabled are Good for Society

1. It is socially irresponsible knowingly to bring an infant with a serious genetic disorder into the world in an era of prenatal diagnosis (37 [37])
   Answers: Strongly Agree (1), Agree (2), Neither Agree nor Disagree (3), Disagree (4), Strongly Disagree (5).
2. The existence of persons with severe disabilities makes society more rich and varied. (37[44])
   Answers: Strongly Agree (1), Agree (2), Neither Agree nor Disagree (3), Disagree (4), Strongly Disagree (5).

(In order for the answers to indicate pro disabled attitudes the numbers of optional answers in question (37[44]) were reversed 1 = 5, 2 = 4 3 = 3 4 = 2 5 = 1).

Pro Parental Autonomy

1. Suppose that a safe and accurate method of pre-conceptional sex selection is developed, such as separation of x and y bearing sperm. Please indicate the extent to which you agree with the following statement. (In the absence of an x-linked disorder): It should be available to everyone, with no restrictions. (29 a 1)
2. Parents should be told all test results relevant to the health of the fetus. (37[1])
3. Parents should be told the sex of the fetus if they ask, at any time during the pregnancy (37[2])
4. Patients are entitled to any service they request and can pay for out-of-pocket. (37[16])
5. Good supportive counseling should be supportive of patient's decisions even if the person counseling disagrees with a decision. (37[17])
6. Withholding any requested service, including prenatal diagnosis, is paternalistic (37[21])
7. Parents should have the right to choose the sex of their children. (37[31]).

Answers: Strongly Agree (1), Agree (2), Neither Agree nor Disagree (3), Disagree (4), Strongly Disagree (5).

(In order for the answers to indicate pro parental autonomy attitudes the numbers of optional answers were reversed 1 = 5, 2 = 4 3 = 3 4 = 2 5 = 1).

Pro Rational Scientific Planning of Life

How would you counsel, in your professional capacity, about termination of pregnancy for each of the following conditions? Assume that a reliable test has been developed. (For each condition choose one for each column.) (in the original 34: 10,11,12,13,14)

1. Familial hypercholesterlemia
2. Predisposition to schizophrenia or bipolar disorder
3. Huntington disease
4. Predisposition to Alzheimer disease
5. Predisposition to Alcoholism

Answers: Not tell them this particular test result + urge parents to carry to term (1), Emphasize positive aspects so they will favor carrying to term without suggesting it directly (2), Try to be as unbiased as possible (3), emphasize negative aspects so they will favor termination without suggesting it directly (4), urge termination (5).

2. A woman should have prenatal diagnosis if medically indicated by her age and family history. (37[4])
3. An important goal of newborn screening is to identify and counsel parental carriers before the next pregnancy (37[7])

4. Before marriage, responsible people should know weather they or their prospective partner carries a genetic disorder that could be transmitted to their children. (37[10])
5. People at high risk for serious disorders should not have children unless they use prenatal diagnosis and selective abortion (37 [19])

Answers: Strongly Agree (1), Agree (2), Neither Agree nor Disagree (3), Disagree (4), Strongly Disagree (5).

(In order for the answers to indicate pro science attitudes the numbers of optional answers were reversed 1 = 5, 2 = 4 3 = 3 4 = 2 5 = 1).

APPENDIX 5

Differences between German and Israeli counselors with regard to social moral dilemmas related to prenatal diagnosis

	Germany Mean (s.d.)	Israel Mean (s.d.)	t-Test
The index			
Pro Selective Abortions	3.02 (0.24)	3.36 (0.25)	−8.980**
Pro Abortion	3.08 (0.85)	3.67 (1.08)	−3.599**
Pro Eugenic practices	2.04 (0.66)	2.02 (0.70)	0.143
Pro Disabled are Good for Society	3.76 (0.63)	2.75(0.82)	8.067**
Pro Parental Autonomy	2.71 (0.44)	3.55 (0.44)	−11.114**
Pro Rational Scientific Planning of Life	2.72 (0.68)	3.36 (0.74)	−5.606**

APPENDIX 6

Interview Questions

1. Personal and professional background
2. Do you think the general public should be required to take any prenatal genetic examinations?
3. Do you think people at high risk for serious genetic disorders should not have children unless they use prenatal diagnosis and selective abortions?
4. To your mind, is there any limit to parental autonomy?
5. What is the purpose of genetic counseling?
6. Please give examples of an especially difficult/easy counseling case, and explain why you have chosen these certain cases.
7. To your mind, what are the biggest hopes and dangers embodied in human genetics?
8. Should there be a limit to men's interference with nature? And if so, where does the line cross?
9. Do you see any conflict between individuals and society, embodied in reproductive genetics? If so, whose rights are more important to you?

10. Do you find it legitimate to abort a fetus on the grounds of its sex?
11. Is it important to you to prevent non-medical abortions?
12. What do you think of your states' abortion law?
13. What is, to your mind, the moral standing of the fetus?
14. Do you believe the disabled make society more rich and varied?
15. Do you think it is fair to the child to bring it into the world with a severe genetic disorder?
16. Is it fair to his siblings?
17. Is it fair to society?
18. Is the principle of non-directivity important to you, and if so, why?
19. Does the threat of legal suits affect your work?
20. Does the history of human genetics influence your practice?
21. Are you satisfied with the availability of genetic counseling in your country?
22. a. For Israelis: Why, to your mind, are Israeli counselors extremely supportive of population screening tests and of selective abortions?
    b. For Germans: Why, to your mind, are German counselors extremely cautious with the application of new genetic reproductive technologies?

APPENDIX 7: THE QUESTIONNAIRE

Dear Genetic Counselors

This is a follow up to Dorothy C. Wertz and John C. Fletcher's survey of geneticists' views on ethics in 37 nations from 1995.

The questionnaire deals with interesting and troubling issues toward which professionals and societies have no clear-cut answers.

I hope you will enjoy your chance to express your opinion on important clinical and public issues. Your views are important to the overall success of the survey.

I will report the results of this study in professional journals and conferences, and I will be glad to talk about it with any of my interviewees.

The completion of the questionnaire should take up to 1 hour.

The questionnaire is strictly anonymous. Information will be reported only in the aggregate.

Please fill in your: (14, 15. 16)

Age ______ years.

Country of Origin ____________

Sex: 1. Male ☐ 2. Female ☐

After you have completed the questionnaire please put it in two envelopes: the first one has no addressee on it, the second is addressed to me, and it has your name on it in order for me to know you have been surveyed. The inner envelopes will be separated and opened up together.

Thank you for your participation,

Yael Hashiloni Dolev

First, we would like some information on your professional activities.

1. Which of the following degrees do you hold? (Check as many as apply).
   a. M.D. (specify)
      0. ☐ Geneticist
      1. ☐ Pediatrician
      2. ☐ Obstetrician/Gynecologist
      3. ☐ General Practitioner
      4. ☐ Other: (specify)____________
   b. 5. ☐ Ph.D., or equivalent (specify field) ___________________
   c. 6. ☐ Other (specify degree and field)_____________________

2. Was your training in medical genetics entirely in your own country ?
   1. Yes ☐
   2. No ☐

   If you received training in other countries, please name

	Length of Training				
	0–1 yr.	1–2 yr.	2–3 yr.	3–4 yr.	>4 yrs.
	1	2	3	4	5
Country					
1.__________	☐	☐	☐	☐	☐
2.__________	☐	☐	☐	☐	☐
3.__________	☐	☐	☐	☐	☐

3. How long have you worked in medical genetics after your training? ________ years(s)

5. Of the time you spend in medical genetics, what percent is

	0–25%	26–50%	51–75%	76–100%
	1	2	3	4
a. related to direct patient care	☐	☐	☐	☐
b. related to patient services (e.g. laboratory tests)	☐	☐	☐	☐
c. research	☐	☐	☐	☐
d. educational activities	☐	☐	☐	☐
e. administration	☐	☐	☐	☐
f. other (specify)____________	☐	☐	☐	☐

11. Where do you conduct *most* of your work?
☐ 1. University-based clinical genetics program
☐ 2. Hospital based clinical genetics program
☐ 3. Independent prenatal diagnosis center
☐ 4. Office: Solo practice
☐ 5. Office: group practice
☐ 6. Hospital owned or managed by a religious organization
☐ 7. Community health clinic
☐ 8. Private commercial laboratory
☐ 9. Other: _______________

12. What percent of your patients have to pay for genetics services from their own funds?
☐ 1. 0
☐ 2. 25%
☐ 3. 26–50%
☐ 4. 51–75%
☐ 5. 76–100%

22. A 36 year old woman has declined chorionic villus sampling (CVS) or amniocentesis. At 23 weeks, she decides to have ultrasound. Ultrasound shows a fetus with a nuchal membrane (excess fold of skin on posterior neck) seen in Trisomy 21, Trisomy 18, or 45,X (Turner Syndrome). She has PUBS (percutaneous umbilical blood sampling), which provides definitive diagnosis. She wants to terminate the pregnancy. If she can not get a legal termination at your institution, what would you do? Please check yes or no for each item and each diagnosis.

If the diagnosis were:

	45,X		Trisomy 21		Trisomy 18
Would you	Yes	No	Yes	No	Yes
Refer her to someone within your own country who would perform the procedure					
Refer her outside the country?					
Misconstrue gestational age and proceed with termination?					
Tell the woman that it is too late for termination and do not offer a referral?					
Tell the woman that she should have come in earlier for prenatal care?					
Tell her that you do not approve of late termination					

27. Below are two cases in which patients have requested prenatal diagnosis. Assume that laws and professional rules permit use of prenatal diagnosis for this purpose in your country). For **each** situation, choose **one** answer.

	How would you Counsel?			
	Provide medical information only	Try to dissuade them from having prenatal diagnosis	Tell patients you will help them achieve their parenting goals	Tell patients that decisions are theirs alone and say you will support whatever decision they make
a. A poor family has five sons. The sixth pregnancy is unwanted, but they say they will make financial sacrifices if the fetus is a girl. Otherwise they will have an abortion				
b. A deaf couple with several hearing children want one who will be deaf like themselves. They both have a type of hereditary deafness that can be prenatally diagnosed. They say they will abort a hearing fetus.				

	What would you do?			If you would perform the test or offer a referral who should pay for the test? (choose one)	
	Perform prenatal diagnosis and provide requested information	Refuse to perform prenatal diagnosis	Refer, including out of country	Patient out-of-pocket only (insurance should not cover)	Public Health Insurance
a. A poor family with 5 sons who want a girl					
b. A deaf couple with hearing children who want a deaf child					

Have you ever counselled in a case similar to A? 1. Yes ☐ 2. No ☐
Have you ever counselled in a case similar to B ? 1. Yes ☐ 2. No ☐

29a. Suppose that a safe and accurate method of pre-conceptional sex selection is developed, such as seperation of X and Y bearing sperm. Please indicate the extent to which you agree or disagree with each statement.

In the absence of an X-linked disorder

	Strongly Agree	Agree	Neither Agree nor Disagree	Disagree	Strongly Disagree
1. It should be available to everyone, with no restrictions					
2. It should be available only to married couples					
3. It should be available only to families with 3 or more children of the same sex					
4. National public health insurance should pay for it					
5. Its use would be morally justified					
6. It would benefit society					
7. I would use it myself					

29b. If a woman is a carrier of an X-linked disorder

	Strongly Agree	Agree	Neither Agree nor Disagree	Disagree	Strongly Disagree
1. I would strongly recommend it					

30. A single blind woman on public welfare has a dominant genetic disorder that leads to legal blindness. She has three sons. Each has a different father. All three fathers are absent. All three sons are 90% blind in spite of early attempts at treatment. The mother tells the geneticist, "Now I think I'll have a little girl." There is a 50% risk that the child will be blind. The geneticist thinks privately, "Don't you think you have brought enough suffering into the world?" For each statement below, indicate the extent to which you agree or disagree.

	Strongly Agree	Agree	Neither Agree nor Disagree	Disagree	Strongly Disagree
1. I Agree with the geneticists private sentiment					
2. The geneticist should convey this sentiment to the woman					
3. The geneticist should try to persuade her not to have another child					
4 The geneticist should suggest sterilization					
5. There should be a law requiring her to be sterilized					

Have you ever counselled in an analogous case? Yes ☐ No ☐

34. How would you counsel , in your professional capacity, about termination of pregnancy for each of the following conditions? Assume that a reliable test has been developed. (For each condition choose one for each column.)

	Column A						Column B	
	Urge parents to carry to term	Emphasize positive aspects so they will favor carrying to term without suggesting it directly	Try to be unbiased as possible	Emphasize negative aspects so they will favor termination without suggesting it directly	Urge termination	Not tell them this particular test result	Would you present carrying to term and placing the child for adoption as an option ?	
							Yes	No
1. The mother's life is in danger								
2. Pregnancy result of rape								
3. Hurler syndrome								
4. Trisomy 13								
5. Cleft lip and palate in a girl								

	Column A						Column B	
	Urge parents to carry to term	Emphasize positive aspects so they will favor carrying to term without suggesting it directly	Try to be unbiased as possible	Emphasize negative aspects so they will favor termination without suggesting it directly	Urge termination	Not tell them this particular test result	Would you present carrying to term and placing the child for adoption as an option ?	
							Yes	No
6. Severe, open spina bifida								
7. Cystic fibrosis								
8. Anenchephaly								
9. Sickle cell anemia								
10. Familial hyperc-holesterlemia (homozygous)								
11. Predisposition to schizophrenia or bipolar disorder								
12. Huntington disease								
13. Predisposition to Alzheimer disease								
14. Predisposition to Alcoholism								
15. 45,X								
16. Trisomy 21								
17. XXY								
18. Cleft lip and palate in a boy								
19. Neurofibromatosis								
20. Severe obesity, in absence of a known genetic syndrome								
21. Achondroplastic dwarfism								
22. Child is not the sex desired by the parents								
23. HIV infection in the fetus								
24. Toxoplasmosis in the fetus in first trimester								
25. Rubella in the fetus in first trimester								
26. PKU								

35. How would *you personally* respond if you yourself in your own life faced the possibility of having a child with the disorders listed below? For **each** disorder, please check **one** box in each section.

	Section A: First 3 months of pregnancy			Section B: 4th to 6th month of pregnancy		
	I would have an abortion	I would **not** have an abortion, but it should be **legal** for others	I would not have an abortion, and it should be **illegal** for others	I would have an abortion	I would **not** have an abortion, but it should be **legal** for others	I would not have an abortion, and it should be **illegal** for others
1. Mother's life is in danger						
2. Pregnancy result of rape						
3. Hurler syndrome						
4. Trisomy 13						
5. Cleft lip and palate in a girl						
6. Severe, open spina bifida						
7. Cystic fibrosis						
8. Anenchephaly						
9. Sickle cell anemia						
10. Familial hyperc-holesterolemia (homozygous)						
11. Predisposition to schizophrenia or bipolar disorder						
12. Huntington disease						
13. Predisposition to Alzheimer disease						
14. Predisposition to alcoholism						
15. 45, X						
16. Trisomy 21						
17. XXY						
18. Cleft lip and palate in a boy						
19. Neurofibromatosis						
20. Severe obesity, in absence of a known genetic syndrome						
21. Achondroplastic dwarfism						
22. Child is not the sex desired by the parents						
23. HIV infection in the fetus						
24. Toxoplasmosis						
25. Rubella in the fetus in first trimester						
26. PKU						

46. The following questions ask about your own satisfaction with different types of situations in genetic counseling.

Please indicate the extent to which you agree or disagree with each statement.

	Strongly Agree	Agree	Neither Agree nor Disagree	Disagree	Strongly Disagree
1. I feel dissatisfied if patients refuse prenatal diagnosis if it is medically indicated.					
2. I am satisfied with a counselling session if the patient gets supportive counselling, even if the patient does not remember the genetic information.					
3. I feel dissatisfied with counselling if patients decide to carry a seriously affected fetus to term					
4. I am satisfied with a counselling session if the patient understands the medical/genetic information provided					
5. I would be dissatisfied if legal abortion were not available after prenatal diagnosis					

37. The following questions are designed to survey your views about a large number of issues. Please indicate the extent to which you agree or disagree with each statement.

	Strongly Agree	Agree	Neither Agree nor Disagree	Disagree	Strongly Disagree
1. Parents should be told all test results relevant to the health of the fetus					
2. Parents should be told the sex of the fetus if they ask, at any time during the pregnancy					
3. Persons who know that they carry a recessive genetic disorder should not mate with another carrier					
4. A woman should have prenatal diagnosis if medically indicated by her age and family history					
5. Society will probably never provide adequate support for people with disabilities.					

(*Continued*)

	Strongly Agree	Agree	Neither Agree nor Disagree	Disagree	Strongly Disagree
6. Maternal serum alpha-feto-protein (MSAFP) screening programs should tell women, **before drawing blood**, that the screening could lead to a decision about abortion.					
7. An important goal of newborn screening is to identify and counsel parental carriers before the next pregnancy.					
8. When patients ask for a procedure that I am unwilling to perform for moral reasons, it is my responsibility to offer a referral, as long as the procedure is legal.					
9. Patients should tell their partners the results of their own genetic tests.					
10. Before marriage, responsible people should know whether they or their prospective partner carries a genetic disorder that could be transmitted to their children.					
11. After taking a test, patients should have the right **not** to know the results.					
12. A woman's decisions about abortion should be her own, without intervention by anyone					
13. Prenatal diagnosis should be provided regardless of the use patients intend to make of it.					
14. An important goal of genetic counselling is to reduce the number of deleterious genes in the population					
15. I do not approve of the abortion of a normal fetus after prenatal diagnosis					
16. Patients are entitled to any service they request and can pay for out-of-pocket.					
17. Good genetic counselling should be supportive of patients' decisions even if the person counselling disagrees with a decision.					
18. governments should require carrier tests for common genetic disorders like cystic fibrosis (CF) and sickle cell anemia.					
19. People at high risk for serious disorders should not have children unless they use prenatal diagnosis and selective abortion.					

	Strongly Agree	Agree	Neither Agree nor Disagree	Disagree	Strongly Disagree
21. withholding any requested service, including prenatal diagnosis, is paternalistic					
23. Some disabilities will never be overcome even with maximum social support.					
26. It is not fair to a child to bring it into the world with a serious genetic disorder.					
27. Parents should not be told the sex of a fetus, except in cases of X-linked disease.					
28. Women are under great social pressure to have prenatal diagnosis					
29. It is not fair to a family's other children knowingly to have a child with a disability.					
30. A woman should have an unqualified right to abortion without viability.					
31. Parents should have the right to choose the sex of their children.					
32. I do not approve of the abortion of a fetus with a genetic defect or malformation, unless the condition leads to mental retardation and death in early childhood.					
34. It is useless to perform diagnostic tests when there is no effective intervention other than abortion.					
37. It is socially irresponsible knowingly to bring an infant with a serious genetic disorder into the world in an era of prenatal diagnosis					
40. There should be a legal option for a woman to carry an anenchephalic fetus to term in order to use the fetus's organs for transplantation.					
41. Consumer groups, including organizations of women and people with disabilities, should be consulted in developing policies about genetic screening and prenatal diagnosis.					
43. If a couple has a child with a disability, the primary responsibility for care usually falls on the mother.					
44. The existence of people with severe disabilities makes society more rich and varied					

19. What is your religious preference ?

☐ 1. Catholic
☐ 2. Protestant, Denomination: ____________
☐ 3. Eastern Orthodox
☐ 4. Jewish
☐ 5. Muslim
☐ 6. Hindu
☐ 7. Buddhist
☐ 8. Other
☐ 9. None

20. How important is religion in your life ?

☐ 1. Extremely important
☐ 2. Very important
☐ 3. Moderately important
☐ 4. Slightly important
☐ 5. Not at all important

21. How often do you attend religious services ?

☐ 1. Never
☐ 2. Less than once a year
☐ 3. Once or twice a year
☐ 4. Several times a year
☐ 5. About once a month
☐ 6. 2–3 times a month
☐ 7. Nearly every week
☐ 8. Several times a week.

BIBLIOGRAPHY

"A Sweet Potato in Dr. Frankenstein's Laboratory". *Haaretz*, 23.06.1999 (in Hebrew). Translated from "The Economist".

Abbott, Alison. Transgenic trials under pressure in Germany. *Nature*. 1996; 380: 94.

Abortion Decision of February 25, 1975, of the Federal Constitutional Court, Federal Republic of Germany, Edmund C. Jann trans. Washington, DC: Library of Congress, 1975.

Abott, Andrew. *The System of Professions: An Essay on the Division of Expert Labor*. Chicago: Chicago University Press, 1988.

Adams, Mark B. *The Well-born Science: Eugenics in Germany, France, Brazil and Russia*, Adams, Mark B (ed.), New York: Oxford University Press, 1990.

Agamben, Giorgio. *Means without End: Notes on Politics*. University of Minnesota Press, 1993.

Agamben, Giorgio. *Homo Sacer: Sovereign Power and Bare Life*. Stanford California: Stanford University Press, 1998.

Alexander, Jeffrey C. Analytic Debates: Understanding the Relative Autonomy of Culture. In: J.C. Alexander and S. Seidman (eds.), *Culture and Society: Contemporary Debates*, Cambridge, UK: Cambridge University Press, 1990, pp. 1–27.

Almog, Oz. *Hatsabar-Dyokan*. (the Sabra: A Portrait) Tel-Aviv: Am Oved (in Hebrew), 1997.

Amir, Delila. *Abortions, a Silenced Issue in Israel. Feminist and International Perspectives*. (in Hebrew), Forthcoming.

Amir, Delila and Biniamin, Orly. Abortion Approval as a Ritual of Symbolic Control. In: C. Feinman, (eds.), *The Criminalization of a Woman's Body*, New York: Harrington Park Press, 1992, pp. 5–16.

Amir, Delila and Navon, David. *The Politics of Abortion in Israel*. Discussion paper series, discussion paper No. 13–89. The Pinhas Sapir Center for Development, Tel-aviv University, Tel-Aviv, Israel (in Hebrew). 1989.

"An offer for a new technology" 16.06.2000, handed to the Israeli health ministry by the Israeli Medical Geneticists Asssociation.

Anderson, Benedict. *Imagined Communities: Reflections on the Origins and Spread of Nationalism*. London: Verso, 1983.

Anson, Jonathan and Meir, Avinoam. Religiosity, Nationalism and Fertility in Israel. *The National Insurance Institute*. 1996; 46: 46–63.

Appadurai, Arjun and Breckenridge, C. Why Public Culture? *Public Culture*. 1988; 1(1): 5–9.

Arendt, Hannah. *The Human Condition*. 2nd ed. Chicago: University of Chicago Press. 1998.

Asch, Adrienne and Geller, Gail. Feminism, Bioethics, and Genetics. In: Susan, M. Wolf (ed.), *Feminism and Bioethics: Beyond Reproduction*. New York and Oxford: Oxford University Press, 1996, pp. 318–345.

Bar, Amos and Rosin, Tali. *The Israeli guide to pregnancy and childbirth*. Tel-Aviv: Zmora-Bitan (in Hebrew), 1998.

Barilan Y. Michael. "Biomedical ethics & Halakha – Approaches to", and "Abortion". The Encyclopedia of Judaism 2nd ed. (J. Neusner et al. ed. Leiden, Brill), Forthcoming.

Barilan, Y. Michael. Know from whence you came, and where you are going: An ethical, historical and social exploration of Jewish bioethics. Chapter 3, Forthcoming.

Bassett, M. William. 'Eugenics and Religious Law: Christianity', In: Warren T. Reich (ed.), *Encyclopedia of Bioethics*. New York: Simon & Schuster Macmillan, 1995, pp. 779–783.

Bauman, Zygmunt. *Modernity and Ambivalence*. Cambridge, 1998.

Beck, Ulrich. *Risk Society*. London: Sage Publications, 1992.

Beck, Ulrich and Beck-Gernsheim, Elizabeth. *The Normal Chaos of Love*. Cambridge: Polity Press, 1995.

Beck-Gernsheim, Elizabeth. "Life as a Planning Project". In: Lash, S., Szerszynski, B. and Wynne, B. (eds.), *Risk, Environment and Modernity: Towards a New Ecology*. London: Sage, 1996, pp. 139–153.

Bekanntmachungen der Bundesärztekammer. Erklärungen zum Schwangerschaftsabbruch nach Pränataldiagnostik, Dt. Ärzteblatt 1998; 95: A-3013–3016 [Heft 47].

Benatouil, Thomas. A tale of two sociologies: The critical and the pragmatic stance in contemporary French sociology. *European Journal of Social Theory*. 1999; 2(3): 379–396.

Berger, Peter and Luckman, Thomas. *The Social Construction of Reality*. New York: Doubleday, 1967.

Bericht der Enquete-Komission. Chancen und Risiken der Gentechnologie, *Deutscher Bundestag*, 10. Wahlperiode, Drucksache 10/6775, January 1987.

Berkovitch, Nitza. Women of Labor: Women and Citizenship in Israel. *Israeli Sociology*. 1999; 2(1): 277–317.

Berufsordnung für die deutschen Ärztinnen und Ärzte, (Muster-)-MBO-Ä 1997 -* in der Fassung der Beschlüsse des 100. Deutschen Ärztetages 1997 in Eisenach geändert durch die Beschlüsse des 103. Deutschen Ärztetages 2 2000 in Köln, published in Deutsches Aerzteblatt.

Biale, David. *Power and Powerlessness in Jewish History*. New york: Schocken Books, 1986.

Birenbaum-Carmeli, Daphna. Our First 'IVF Baby': Israel and Canada's Press Coverage of Procreative Technology. *International Journal of Sociology and Social Policy*. 2000; 20/7, pp. 1–38.

Birenbaum-Carmeli Daphna. Prevalence of Jews as subjects in genetic research: Figures, explanation, and potential implications. *American Journal of Medical Genetics*. 2004; 130A (1): 76–83.

Blech, J.D. Abortion and the Jewish Law. In: Thomas. W. Hilgers, Dennis J. Horan, and David, Mall (eds.), *New Perspectives on Human Abortion*, Fredrik: Athenia Books, University Publications of America, 1981.

Bloor, David. *Knowledge and Social Imagery*. Chicago: University of Chicago Press, 1991.

Blossfeld, Hans-Peter and Rohwer, Goetz. West Germany. In: Hans-Peter Blossfeld. Boulder (ed.), *The New Role of Women: Family Formation in Modern Societies*. Westview Press. NY. 1995, pp. 57–76.

Blumberg, Lisa. The Politics of Prenatal Testing and Selective Abortion. *Sexuality and Disability*. 1994; 12(2): 135–153.

Bock, Gisela. Racism and Sexism in Nazi Germany: Motherhood, Compulsory Sterilization, and the State. In: Bridenthal, R. Grossman, A and Kaplan, M. (eds.), *When Biology Became Destiny*. New York: Monthly Review Press, 1984, pp. 271–295.

Bock, Gisela. Antinatalism, Maternity, and Paternity in National Socialist Racism. In: Bock, G. and Thane, P. (eds.), *Maternity and the Rise of European Welfare Stats, 1880s–1950s*. London: Routledge, 1991, pp. 233–255.

Boltanski, Luc and Thévenot, Laurent. The Sociology of Critical Capacity. *European Journal of Social Theory*. 1999; 2(3): 359–377.

Boltanski, Luc and Thévenot, Laurent. *On Justification: Economies of Worth*. Princeton University Press, Princeton 2006.

Bordo, Susan. Feminism, Foucault and the Politics of the Body. In: Janet Price and Margrit Shildrick (eds.), *Feminist Theory and the Body: A Reader*. Edinburgh: Edinburgh University Press, 1999, pp. 246–257

Bourdieu, Pierre. *Distinction : A Social Critique of the Judgment of Taste*. Cambridge, MA: Harvard University Press, 1984.

Bourdieu, Pierre. *Outline of a Theory of Practice*. Trans. Richard Nice. Cambridge MA: Cambridge University Press, 1977.

"Bush signed a controversial law forbidding late term abortions". *Haaretz*, 07.11.2003 (in Hebrew). (No reporter mentioned)

Cahille, Lisa Sowle 'Abortion, Religious Tradition: Roman Catholic Perspectives', In: Warren T. Reich (ed.), *Bioethics: Sex, Genetics & Human Reproduction*. New York: Simon and Schuster. Macmillan Compendium: Single Volume Encyclopedia, 1998, pp. 34–37

Canguilhem, Georges. *The Normal and the Pathological*. New York: Zone Books, (1991)1904.

Caplan, Arthur. L. Neutrality is not Morality: The Ethics of Genetic Counseling. In: Bartels, M. Dianne, LeRoy, S. Bonnie and Caplan, L. Arthur (eds.), *Prescribing Our Future: Ethical Challenges in Genetic Counseling*. Aldine de Gruyter, New York, 1993, pp. 149–165.

Carron Sher. et al. Factors Affecting Performance of Prenatal Genetic Testing by Israeli Jewish Women. *American Journal of Medical Genetics*. 2003; 120A: 418–422.

Caspi Arye "Genetic Defect". *Haaretz Magazine*, 7.09.2001 (in Hebrew).

Charo, Alta R. and Rothenberg, Karen H. The Good Mother: The Limits of Reproductive Accountability and Genetic Choice. In: Karen, H. Rothenberg and Elizabeth, J. Thomson (eds.), *Women and Prenatal Testing*. Ohio State University Press, Columbus, 1994, pp. 105–129.

Chemke, Juan M and Steinberg, A. Ethics and Medical Genetics in Israel. In: Wertz, D.C and Fletcher, J.C. (eds.), *Ethics and Human Genetics: A cross Cultural Perspective*. Berlin: Springer-Verlag, 1989, pp. 271–284.

Chemke, Juan M and Zlotogora, Joël. Genetic Services in Israel. *European Journal of Human Genetics*. 1997; 5(2): 105–111.

Chen, Shoshana. "The Rabbi said: Marriage is forbidden (without genetic test)" *Yediot Aharonot*, 19.08.2001 (in Hebrew).

C.I.A World Fact Book. 2001. C.I.A.gov/cia/publications/factbook.

Clarke, Angus. Is non-directive genetic counseling possible? *The Lancet*. 1991; 338(October 19): 998–1001.

Cohen, J. Maurie, Science and Society in Historical Perspective: Implications for Social Theories of Risk. *Environmental Values*. 1999; 8: 153–176.

Cohen, Pamela. E., Wertz, Dorothy. C., Nippert, Irmgard, and Wolff, Gerhard. Genetic Counseling Practices in Germany: A Comparison between East German and West German Geneticists. *Journal of Genetic Counseling*. 1997; 6(1): 61–80.

Cohen, Tirza. The history and development of human genetics in Israel. In: Dronamraju K.R. (ed.), *The History and Development of Human Genetics*. World scientific, 1992, pp. 174–184.

Cohen-Almagor Rafi, and Snir, Yehuda. The decision-making process of abortion high committees. *Harefuah*. 2000; 138(12): 1009–1014 (in Hebrew).

Cohen-Almagor Rafi and Shmueli, Merav. Can life be evaluated? The Jewish Halachic approach vs. the quality of life approach in medical ethics: A critical view. *Theoretical Medicine and Bioethics*. 2000; 21: 117–137.

Conklin B.A. and Morgan, L.M. Babies, bodies, and the production of personhood in North America and a native Amazonian society. *Ethos*. 1996; 24(4): 657–694.

Connery, John S. *Abortion: The Development of the Roman Catholic Perspective*. Chicago: Loyola University press, 1977.

Coucil of the Protestant Church in Germany: Der Schutz menschlicher Embryonen darf nicht eingeschränkt werden, 22. May 2001 http://www.ekd.de/presse/39/ 5220.html retrieved 24.09.2001.

Crane, Diana. *The Sanctity of Social Life*. New York: Russell-Sage, 1975.

Davis, Nanette. *From Crime to Choice: The Transformation of Abortion in America*. Westport, Connecticut: Greenwood Press, 1985.

Davison, Hunter J. and Davis, J.E. Cultural Politics at the Edge of Life. In: Critchlow, D.T. (ed.), *The Politics of Abortion and Birth Control in Historical Perspective*. Pennsylvania: The Pennsylvania State University Press, 1996, pp. 103–128.

Dean, Mitchell. Four Theses on the Powers of Life and Death. *Contretempts*. 2004; 5, December: 16–29.

Deichmann, Ute. *Biologists under Hitler*. Cambridge, Massachusetts: Harvard University Press, 1996.

Deichmann, Ute. Emigration, isolation and the slow start of molecular biology in Germany, *Studies in the History and Philosophy of Biological and Biomedical Sciences* 2002; 33: 433–455.

Dietrich, Donald, J. Catholic eugenics in Germany, 1920–1945: Hermann Muckermann, S.J. and Joseph Mayer. *Journal of Church and State*. 1992; 34(3): 575–601.

Dinkel, R.H. and Lebok, U.H. The Fertility of Migrants Before and After Crossing the Border: The Ethnic German Population from the Former Soviet Union as a Case Study. *International Migration Journal*. 1997; 35(2): 253–270.

Dixon, Bernard. Who's who in European antibiotech. *Bio/Technology*. 1993; 11: 45.

Dorbritz, Juergen. Bericht 1994 ueber die demographische Lage in Deutschland. *Zeitschrift fuer Bevoelkkerungswissenshaft* 1993–4; 19(4): 393–473.

Dorbritz, J. and Schwarzer, K. Kinderlosigkeit in Deutschland – ein Massenphaenomen? Analysen zu Erscheinungsformen und Ursach. *Zeitschrift fuer Bevoelkerungswissenschaft*. 1996; 3: 231–261.

Douglas, Mary. *Purity and Danger: An Analysis of Concepts of Pollution and Taboo*. London: Routledge and Kegan Paul, 1966.

Douglas, Mary. *Natural Symbols. Exploration in Cosmology*. Harmodsworth: Penguin, 1970.

Douglas, Mary. *Risk Acceptability According to the Social Sciences*. London: Routledge and Kegan Paul, 1986.

Douglas, Mary. Risk as a forensic resource. *Daedalus*. 1990; 119: 1–16.

Douglas, Mary. *Risk and Blame: Essays in Cultural Theory*. London: Routledge, 1992.

Douglas, Mary and Wildavsky, Aaron. *Risk and Culture: An Essay on the Selection of Technological and Environmental Dangers*. Berkley: University of California Press, 1982.

Dror, Yuval. "Israel is for cloning, but it is also against it" *Haaretz*, 12. 12.2003 (in Hebrew).

Dunne, Cara and Warren, Catherine. Lethal autonomy: The malfunction of the informed consent mechanism within the context of prenatal diagnosis of genetic variants. *Issues in Law & Medicine*. 1998; 14(2 Fall): 165–203.

Duster, Troy. *Backdoor to Eugenics*. New York: Routledge, 1990.

Eberstadt, Nicholas. Demographic shocks after Communism: Eastern Germany, 1989–1993. *Population and Development Review*. 1994; 20(1): 137–152.

Embryo Protection Law: Gesetz zum Schutz von Embryonen (Embryonenschutzgesetz – EschG), in the version of Dec. 13th 1990. BGBI. I. S. 2747.

Erikson, Susan. Maternity Care Policies and Maternity Care Practices: A Tale of Two Germany's. In: Raymond De Vries et al. (eds.), *Birth by Design: Pregnancy, Maternity Care and Midwifery in North America and Europe*. New York: Routledge, 2001, pp. 203–217.

Erikson, Susan. Post-diagnostic abortion in Germany: Reproduction gone awry, again? *Social Science & Medicine*. 2003; 56: 1987–2001.

Erklärungen zum Schwangerschaftsabbruch nach Pränataldiagnostik, Bekanntmachungen der Bundesärztekammer, Dt Ärzteblatt 1998; 95: A-3013-3016 [Heft 47].

Ettorre, Elizabeth. Experts as 'storytellers' in reproductive genetics: Exploring key issues. *Sociology of Health and Illness*. 1999; 21(5): 539–559.

Falk, Rafael. Zionist Settlement as a Eugenic Enterprise. *Alpayim*. 2002; 23: 179–199 (in Hebrew).

Featherstone, Mike. *Undoing Culture: Globalization, Postmodernism and Identity*. London: Sage, 1995.

Feldman, David. M. *Health and Medicine in Jewish Tradition*. New York: Crossroad, 1986.

Feldman, David. M. Eugenics and Religious Law: Judaism. In: Warren T. Reich (ed.), *Bioethics: Sex, Genetics & human reproduction*. New York: Simon and Schuster. Macmillan Compendium: Single Volume Encyclopedia, 1998.

Felstiner, William. L.F. and Abel, Richard. L. and Sarat, Austin. The Emergence and Transformation of Disputes: Naming, Blaming, Claiming… *Law and Society Review*. 1980–1; 15: 3–4.

Fiedler, Leslie. *Tyranny of the Normal: Essays on Bioethics, Theology and Myth*. Massachusetts: David R. Godine, Publisher, Inc, 1996.

Filc, Daniel. Health as Merchandise: Public Health in Israel: Between the State, Civil Society and the Market. In: Yoav Peled and Adi Ophir (eds.), *Israel: From Mobilized to Civil Society?* Tel-Aviv: The Van Leer Jerusalem Institute/Hakkibutz Hameuhad Publishing, 2001 (in Hebrew).

Fine, Beth E. The Evolution of Nondirectiveness in Genetic Counseling and Implications of the Human Genome Project. In: Bartels, M. Dianne, LeRoy, S. Bonnie and Caplan, L. Arthur (eds.), *Prescribing Our Future: Ethical Challenges in Genetic Counseling*. Aldine de Gruyter, New York, 1993, pp. 101–117.

Fink, Daniel. and Glick, Simon. Misinformation Among Physicians About Dangers of Fetal X-ray Exposure. *Harefuah*. 1993; 124(11): 717–719 (in Hebrew).

Fischer, Claude, S. *America Calling. A Social History of the Telephone to 1940*. Berkley: University of California Press, 1992.

Fogiel-Bijaoui, Sylvia. Families in Israel: Between Familism and Post-Modernity. In: Dafna, N. Israeli et al. (eds.), *Sex Gender Politics: Women in Israel.* Tel Aviv: Hakibbutz Hameuchad, 1999, pp. 107–166 (in Hebrew).

Foucault, Michel. *The History of Sexuality, vol. 1: An Introduction.* NY: Pantheon, 1978.

Foucault, Michel. Governmentality. In: Burchell, G, Gordon, C. and Miller, P. (eds.), *The Foucault Effect: Studies in Governmentality.* Chicago: The University of Chicago Press, 1991, pp. 87–104.

Fox, Renee. Ethical and Existential Developments in Contemporary Medicine. *Milbank Memorial Fund Quarterly* 1974; 54: 445–483.

Fox Keller, Evelyn. *Reflections on Gender and Science.* New Haven: Yale University Press, 1985, pp. 75–95.

Fraser, F. Clarke. Genetic Counseling. *American Journal of Human Genetics.* 1974; 26: 636–661.

Freilich, Rivka. "87% of the religious women would not abort a fetus which will be mentally retarded". *Yediot Ahront.* 1.12.2003 (in Hebrew).

Friedlander, Henry. *The Origins of Nazi Genocide: From Euthanasia to the final Solution.* Chapel Hill: The University of North California Press, 1995.

Friedlender and Goldschneider, C. *The Population of Israel.* New York: Columbia University Press, 1979.

Funk, Nanette. Abortion and German Unification. In: Funk, N & Mueller, M. (eds.), *Gender Politics and Post-Communism: Reflections from Eastern Europe and the former Soviet Union.* New York, London: Routledge, 1993, pp. 194–200.

Galligan, D.J. Introduction. In: D.J. Galligan (eds.), *Socio-Legal Studies in Context: The Oxford Center Past and Future.* Blackwell publishing, 1995.

Gauthier, A.H. The Western European Governments attitudes and responses to the demographic and family questions. *Paper presented at the European Science Foundation Coference, St. Martin,* 1992.

Geertz, Clifford. *Interpretation of Cultures: Selected Essays.* New York: Basic Books, 1973.

German Abortion law 1995, paragraph 218 a part 2. Translated from German in Wegner, R-D, and Becker, R. 1997. Prenatal diagnosis in Germany. *European Journal of Human Genetics.* 5(1): 32–38.

German National Ethics Council (Nationaler Ethikrat), 2003. "Genetic diagnosis before and during pregnancy. Opinion". Available online at: http://www.ethikrat.org/_english/publications/Stn_PID_engl.pdf.

Giddens, Anthony. *Modernity and Self-identity: Self and Society in Late Modern Age.* Cambridge: Polity Press, 1991.

Gilligan, Carol. *In a Different Voice, Psychological Theory and Women's Development.* Cambridge, Mass: Harvard University Press, 1982.

Ginsberg, Gary. et al. Cost Benefit Analysis of a National Screening Programme for Cystic Fibrosis in an Israeli Population. *Health Economics.* 1994; 3: 5–23.

Glendon, Marry Ann. *Abortion and Divorce in Western Law: American failures, European Challenges.* London and Cambridge Massachusetts: Harvard University Press, 1987.

Goffmann, Erving. *Frame Analysis.* Philadelphia: University of Pennsylvania Press, 1975.

Golan, Tal. Introduction. *Israel Studies.* 2004; 9(2): v–viii.

Goldberg, Albert I. The Relevance of Cosmopolitan/Local Orientations to Professional Values and Behavior. *Sociology of Work and Occupations.* 1976 3(3): 331–355.

Goldsmith, E. Ronen, and Ronen, I. Changing Marriage Systems in the Jewish Community of Israel. *Annual Human Genetics.* 1960; 24: 191–195. Quoted in Zlotogora, Joel. and Chemke, Juan. Medical Genetics in Israel. *European Journal of Human Genetics.* 1995; 3:147–154.

Good, Byron. *Medicine, Rationality and Experience: An Anthropological Perspective.* Cambridge: Cambridge University Press, (1994)1997.

Goode, William. Community within a Community. *American Sociological Review.* 1957; 22: 194–200.

Goode, William. Professions and non-professions. In: Howard, M. Vollmer and Donald L. Mills (eds.), *Professionalization.* Engelwood Cliffs, NJ: Prentice-Hall, 1966, pp. 33–45.

Goodman Richard.M. et al. Medical genetics in Israel. *Journal of Medical Genetics.* 1989; 26: 179–189.

Gooldin, Sigal. Anorexia in Israel or 'Israeli Anorexia'? Some Remarks on a Culture-Bound Syndrome in a Glocal Context. *Israeli Sociology.* 2002; 4(1): 105–142.

Gooldin, Sigal. Technologies of Happiness: Fertility Management in a Welfare Pro-Natal State. In: A. Kemp and Y. Yona (eds.), *Citizenship Gaps: Migration, Fertility and Identity*. Jerusalem: Van Leer Institute–Hakibbutz Hameuchad, (2007) (in Hebrew).

Gross, Michael L. After Feticide: Coping with Late-Term Abortion in Israel, Western Europe, and the United States. *Cambridge Quarterly of Health Care Ethics*. 1999; 8(4): 449–462.

Gross, Michael L. Ethics, Policy and Rare Genetic Disorders: The Case of Gauche Disease in Israel. *Theoretical Medicine*. 2002; 23: 151–170.

Grossman, Anita. The Debate that Will Not End: The Politics of Abortion in Germany from Weimar to National Socialism and the Postwar Period. In: Berg, M. and Cocks, G. (eds.), *Medicine and Modernity. Public Health and Medical Care in 19th and 20th century Germany*. Washington: German Historical Institute and Cambridge University Press, 1997, pp. 193–211.

Gruendel Johannes. Theologisch-Etische Aspekte zur praenatalen Diagnostik. In: Murken, J. (ed.), *Prenatale Diagnostik und Therapie*. Stuttgart: Enke Verlag. 1987, p. 314.

Habermas, Jürgen. *Toward a Rational Society*. London and Edingurgh: Heinemann (1968)1971.

Habermas Jürgen. *Wahrheit und Rechtfertigung*. Frankfurt/Main 1999, pp. 271–318.

Habermas, Jürgen. On the way to liberal eugenics? The dispute over the ethical self-understanding of the species. *Paper from colloquium in Law, Philosophy and Political Theory*. New York University, August 2001.

Habermas, Jürgen. *The Future of Human Nature*. Cambridge, UK: Polity Press, 2003.

Hacking, Ian. *The Taming of Chance*. Cambridge: Cambridge University Press, 1990.

Hadley, Janet. Prenatal Tests: Blessings and Burdens. In: Ellie Lee (ed.), *Abortion Law and Politics Today*. London: Mcmillan Press Ltd, 1998, pp. 172–183.

Haelyon, Hilla. *The Experience of Israeli Women Undergoing In Vitro Fertilization Treatments. A Combined Model for the Examination of Body-Identity-Emotion Triangle*. Ph.D. Dissertation. Bar-Ilan University, Department of Sociology and Anthropology, 2004.

Hammer, M.F. Redd, A.J. Wood, E.T. Bonner, M.R. Jarjanazi, H. Karafet, T. Santachiara-Benerecetti, S. Oppenheim, A. Jobling, M.A. Jenkins, T. Ostrer, H. Bonne-Tamir, B. Jewish and Middle Eastern Non-Jewish Populations Share a Common Pool of Y-Chromosome Biallelic Haplotypes, Proceedings of the National Academy of Science USA 2000; 97(12): 6769–6677.

Harrer, Herbert. Aspects of Failed Family Planning in the United States of America and Germany. *The Journal of Legal Medicine*. 1994; 15: 89–127.

Harris, R. and Reid, M. Medical Genetic Services in 31 Countries: An Overview. *European Journal of Human Genetics*. 1997; 5(2): 3–21.

Hashiloni-Dolev, Yael. Who is a Wanted Child? Selective Abortions Due to Sex-Chromosome Anomalies in Israel and Germany. *Theory and Criticism*. 2004; 25: 97–123.

Hazan Haim. Between the Times: The Body of Israeli identity. *Zmanim*. 1999; 17–18: 68–69 (in Hebrew).

Hazan, Haim. The Body Sociological – On the Status of Somatic Representations. *Israeli Sociology*. 2003; 5(1): pp. 219–230 (in Hebrew).

Health Condition in Israel. 1999. Health Ministry, the National Center for Monitoring Disease (in Hebrew).

Herf, Jeffrey. *Divided Memory: The Nazi Past in the two Germanys*. Cambridge, MA: Harvard University Press, 1997.

Herzog, Dagmar. Pleasure, Sex and Politics Belong Together: Post-Holocaust Memory and the Sexual Revolution in West Germany. *Critical Inquiry*, 1998; 24 (Winter): 393–444.

Heyd, David. Are 'Wrongful Life' Claims Philosophically Valid? *Israel Law Review*. 1986; 21: 574–590.

Heyd, David. *Genethics: Moral Issues in the Creation of People*. Berkeley: University of California Press, 1992.

Heyd, David. Prenatal diagnosis: Whose right? *Journal of Medical Ethics*. 1995; 21: 292–297.

Heyd, David. The right to be born free of birth defects. In: Raphael Cohen-Almagor (ed.), *Moral Dilemmas in Medicine*. Van-Leer Jerusalem Institute/Hakibutz Hameuchad Publishing House, 2002, pp. 255–261 (in Hebrew).

Hirsch, Dafna. We are here to bring the West: Hygiene education within the Jewish community of Mandatory Palestine. Ph.D. Thesis. Tel-Aviv University, (Forthcoming).

Inge, W.R. Eugenics and Religion. *Eugenics Review*. (1921)1968; 60 (2 January): 92–98.

Israel National Bioethics Council-guidelines for PGD in Israel, 2005.

Israeli Law of Genetic Information (2000) (in Hebrew).

Ivry, Tsipy. *Pregnant with Meaning: Conceptions of Pregnancy in Japan and Israel*. Doctoral Dissertation submitted to the Department of Sociology and Anthropology, Hebrew University of Jerusalem, 2004.

Izhar, Erez, A Matter of Genetic. *Ha'ikar haberiyut: Kupat Holim Meuhedet Quarterly*. 2004; 22: 28–32 (in Hebrew).

Jaber L, Baley-Wilson JE, Haj-Yehia, M. Hernandez, J. Shohat, M. Consanguineous Matings in an Israeli-Arab community. *Arch Pediatr Adolesc Med* 1994; 148: 412–415. Quoted in Zlotogora, Joel. and Chemke, Juan. Medical Genetics in Israel. *European Journal of Human Genetics*. 1995; 3: 147–154.

Jakobovits, Immanuel. *Jewish Medical Ethics: A Comparative and Historical Study of the Jewish Religious Attitude to Medcine and its Practices*. New York: Bloch Publishing Company, [1959]1967, pp. 153–169.

Jarausch, Konrad.H. *The Rush to German Unity*. Oxford University Press New York – USA, 1994.

Jellinek, Samuel. *Wrongful Life: Rights of Claim and Compensation*. Tel Aviv: Eshley Publishing house, 1997 (in Hebrew).

Kaback, M. Lim-Steele, J. Dabholkar, D. Brown, D. Levy, N. Zeiger, K. Tay-Sachs Disease-Carrier Screening, Prenatal Diagnosis, and the Molecular Era: An International Perspective 1970 to 1993. *The International TSD Data Collection Network. JAMA* 1993; 270: 2307–2315.

Kahn, Susan Martha. *Reproducing Jews: A Cultural Account of Assisted Conception in Israel*. Durham and London: Duke University Press, 2000.

Kanaaneh, Rhoda, A. *Birthing the Nation: Strategies of Palestinian Women in Israel*. Berkeley: University of California Press, 2002.

Katz–Rothman, Barbara. *The Tentative pregnancy: How Amniocentesis Changes the Experience of Motherhood*. New York: W.W. Norton & Company, 1986.

Keller, Heidi, Zach, Ulrike and Abels, Monika. The German Family: Families in Germany. In: Jaipaul, L. Roopnarine & Uwe P. Gielen (eds.), *Families in Global Perspective*. Boston MA: Pearson Allyn and Bacon, 2005, pp. 242–258.

Kersting-Wilmsmeyer, Andreas. Views of the German Cystic Fibrosis Association, quoted in chapter 3: Country report from Germany In: Nippert, Irmgard., Clausen, Hans., Frets, Petra., Niermeijer Martinus F and Modell Michael (eds.), *Evaluating Cystic Fibrosis Carrier Screening Development in Northern Europe: Denmark, The Federal Republic of Germany, the Netherlands, and the United Kingdom*, Münster: Westf. Wilhelms-Universität, Women's Health Research Series 1998, pp. 43–70.

Kaczor, Christopher. *The Edge of Life: Human Dignity and Contemporary Bioethics*. Dordrecht: Springer, 2005, pp. 23–25.

Kevles, Daniel J. *In the Name of Eugenics: Genetics and the Uses of Human Heredity*. New York: Knopf, 1985.

Kimor, Dalit. "A Tailored Baby" A documentary film presented on the Israeli channel 1, public television, 2001.

Kirsh, Nurit. Population Genetics in Israel in the 1950s: The Unconscious Internalization of Ideology. Isis. 2003; 94: 631–655.

Kirsh, Nurit. Genetic Research on Israel's Populations: Two Opposite Tendencies. In: Volker Roelcke, Giovanni Maio (eds.), *Twentieth Century Ethics of Human Subjects Research: Historical Perspectives on Values, Practices, and Regulation*. Stuttgart: Steiner, 2004, pp. 309–317.

Kish, Shifra. *"Deaf Discourse" the social construction of the deafness in a Bedouin community*. MA dissertation, TA University, 2000 (in Hebrew).

Kleinman, Arthur. *Writing at the Margin: Discourse between Anthropology and Medicine*. Berkley: University of California Press, 1995.

Kline, R.R. Technological Determinism. In: Smelser, Neil, J. and Baltes, Paul B. (eds.), *International Encyclopedia of the Social and Behavioral Sciences*, Oxford: Elsevier Science Ltd. 2001, pp. 15495–15498.

Knorr-Cetina, K.D. (ed.), *Science in Context*. Sage: London, 1982.

Koch, Lene. The Meaning of Eugenics: Reflections on the Government of Genetic Knowledge in the Past and the Present. *Science in Context*. 2004; 17(3): 315–331.

Koonz, Claudia. *Mother's in the Fatherland: Women, the Family and Nazi Politics*. New York: St. Martins, 1986.

Kraus, Elliott, A. *Death of the Guilds: Professions, States, and the Advance of Capitalism 1930 to the present*. New Haven: Yale University Press, 1996.

Krones Tanja. Public, Expert and Patients Opinions on Preimplantaion Genetic Diagnosis (PGD) in Germany. *Reproductive Biomedicine Online*. 2005; 10(1): 116–123 Online: www.rbmonline.com/Article/1547 on web 10 November 2004.

Krones, Tanja and Richter, Gerhard. Preimplantation genetic diagnosis: European perspectives and the German situation. *Journal of Medicine and Philosophy*. 2004; 29: 623–640.

Krones Tanja et al. What is the preimplantation embryo? *Social Science and Medicine*. 2006; 63: 1–20.

Lamont, Michele. *Money, Morals, and Manners: The Culture of French and American Upper-Middle Class*. Chicago and London: The University of Chicago Press, 1992.

Landau, Ruth. Assisted Reproduction in Israel and Sweden: Parenthood at any Price? *International Journal of Sociology and Social Policy*. 1996; 16(3): 29–47.

Lanzerath, Dirk. Bioethics in Germany: Debates and infrastructure. *Ann Ist Super Sanità*. 2004; 40(3): 287–296.

Latour, Bruno. *Science in Action: How to follow Scientists and Engineers through Society*. Cambridge, MA : Harvard University Press, 1987.

Latour, Bruno. *We have never Been Modern*. Cambridge: Harvard University Press, 1993.

Latour, Bruno. Is Re-modernization Occuring- And if so, How to Prove It? A Commentary on Ulrich Beck. *Theory, Culture & Society*. 2003; 20(2): 35–48.

Leah Dayan et al. v. Clalit Health Services, C.C. 1037/93, 9.12.98, D.C.

Lee, Ellie. Tensions in the Regulation of Abortion in Britain. *Journal of Law and Society*. 2003; 30(4): 532–553.

Legal Pregnancy Terminations: Trends in the activities of the abortion comities 1990–1999, Socio-demographic characteristics of women who appeal to abortion comities, 1994–6. September 2000. Jerusalem: Israel's health ministry (in Hebrew).

Legge, Jerome, S. Predictions of Abortion Attitudes in the Federal Republic of Germany. *The Journal of Politics*. 1983; 45: 759–766.

Leitlinien zur Erbringung humangenetischer Leistungen: 1. Leitlinien zur Genetischen Beratung. *Med Genetik*. 8 vol. 3, Supplement pp. 1–2. Berufsverband Medizinische Genetik e.V. 1996.

Linke, Uli. *German Bodies: Race and Representation after Hitler*. New York: Routledge, 1999.

Lippman, Abby. Prenatal Genetic Testing and Screening: Constructing Needs and Reinforcing Inequities. *American Journal of Law and Medicine*. 1991; XVII (1&2): 15–50.

Lock, Margaret. The Politics of Mid-Life and Menopause: Ideologies for the Second Sex in North America and Japan. In: Shirley Lindenbaum and Margaret Lock (eds.), *Knowledge, Power and Practice: The Anthropology of Medicine and Everyday Life*. Berkeley: University of California Press, 1993, pp. 330–363.

Lock, Margaret. *Twice Dead: Organ Transplants and the Reinvention of Death*. Berkeley: University of California Press, 2002.

Lori, Aviva. "What did Daddy Mean?" *Haaretz*, April 16, 2003 (in Hebrew).

Luescen, G., Niemann, S., and Apelt, P. The Integration of Two Health Systems: Social Stratification, Work and Health in East and West Germany. *Social Science and Medicine*. 1997; 44(6): 883–899.

Luker, K. *Abortion and the Politics of Motherhood*. Berkeley and Los Angeles, California: University of California Press, 1985.

Lupton, Deborah. *Risk*. London: Routledge, 1999.

Maleck-Lewy, Eva and Marx Ferree, Myra. Talking about Women and Wombs: The discourse of Abortion and Reproductive rights in the G.D.R. during and after the Wende. In: Gal, Susan and Kligman, Gail (eds.), *Reproducing Gender: Politics, Publics and Everyday Life after Socialism*. Princeton: Princeton University Press, 2000, pp. 93–117.

Martin, Emily. The Egg and Sperm: How Science has Constructed a Romance Based on Stereotypical Male-Female Roles. Journal of Women. *Culture and Society*. 1991; 16(31): 485–501.

Marx, Zvi. C. *Disability in Jewish Law*. London: Routledge, 2002.

Marx-Ferree, Myra, Gamson., Willian Anthony, Gerhards Juergen and Rucht Dieter. *Shaping Abortion Discourse: Democracy and the Public Sphere in Germany and the United States*. Cambridge University Press, 2002.

Mason, J.K. *Medico-Legal Aspects of Reproduction and Parenthood*. Dartmouth: Ashgate, 1998.

Melamed, Shoham. *The Janus Face of the 'Demographic Threat': Gender, Ethnicity, Nationalism and the Politics of Fertility in Israel in the 1950's*. MA Thesis, Department of Sociology and Anthropology, Tel-Aviv University, 2002 (in Hebrew).

Merchant, Jennifer. Confronting the consequences of Medical Technology: Policy frontiers in the US & France. In: Githens Marianne and Mcbride Stetson Dorothy (eds.), *Abortion Politics: Public Policy in Cross-Cultural Perspective*. New York and London: Routledge, 1996, Chapter 9.

Michael, Adrian "A Philosophy of Embryo Use". FAZ 1.07.01.

Milbank, John. *Theology and Social Theory: Beyond Secular Reason*. Oxford: Blackwell publishers, 1990.

Mishori-Dery, Anat, Shoham-Vardi Ilana and Carmi, Rivka. *Different Perceptions and Attitudes Regarding Prenatal Testing Among Service Providers and Consumers*. Forthcoming.

Moeller, Robert, G. *Protecting Motherhood: Women and the Family in the Politics of Postwar West Germany*. Berkeley: University of California Press, 1993.

Moore, W.E. The professions and social structure. In: Talcott Parsons (ed.). *Essays in Sociological Theory*. Glencoe, IL, Free Press, 1970, pp. 34–49.

Morag-Levine, Noga. Abortion in Israel: Community, Rights and the Context of compromise. *Law and Social Inquiry* 1994; 19(2 Spring): 313–335

Morgan, Lynn, M. "Life Begins When They Steal Your Bicycle": Cross-Cultural Practices of Personhood at the Beginnings and Ends of Life. Spring 2006; 34: 8–15.

Morley, David. Where the Global Meets the Local: Notes from the Sitting Room. *Screens*. 1991; 32(1): 1–15.

Morris, Douglas G. Abortion and Liberalism: A Comparison between the Abortion Decisions of the Supreme Court of the United States and the Constitutional Court of West Germany. *Hastings International and comparative law review*. 1998; 11: 159–245.

Müller-Hill, Benno. Genetics after Auschwitz. *Holocaust and Genocide Studies*. 1987; 2(1): 3–20.

Müller-Hill, Benno. *Murderous Science : Elimination by selection of Jews, Gypsies, and Others, Germany 1933–1945*. Oxford: Oxford University Press, (1984)1988.

Müller-Hill, Benno. The amnesia of human geneticists. For: Hippocrates Betrayed. Medicine in the "Third Reich". Conference organized by the New York Academy of Science, and held at the US Holocaust Memorial Museum, Washington, DC, January 24th, 1996.

Mushaben, Joyce, Giles, Geoffrey and Lennox, Sara. Women, Men and Unification: Gender Politics and the Abortion Struggle Since 1989. In: Jarausch, K.H. (ed.), *After Unity: Reconfiguring German Identities*. Berghahn Books, 1997, pp. 137–172.

Navon, Liora. *Lepers as Figurative Types: A Social Biography of Metaphor*. Ph.D. Thesis: Tel-Aviv University, 1991 (in Hebrew).

Nikolas, Rose. The Politics of Life Itself. *Theory, Culture and Society*. 2001; 18/6: 1–30.

Nippert, Irmgard. History of Prenatal Genetic Diagnosis in the Federal Republic of Germany. In: Reid, M. (ed.), *The Diffusion of Four Prenatal Screening Tests Across Europe*. London: King's Fund Centre for Health Services Development, 1992a, pp. 48–69.

Nippert, Irmgard. Provision of Prenatal Genetic Diagnosis in the Federal Republic of Germany. In: Kuliev, A. Greendale, K. Penchaszadeh, V and Paul. W.N (eds.), *Genetic Services Provision: An International Perspective*. New York: White Plains, March iof Dimes Birth Defects Foundation, 1992b, pp. 68–81.

Nippert, I. Horst, J. and Schmidke, J. Genetic Services in Germany. *European Journal of Human Genetics*. 1997; 5(2): 81–88.

Nippert, Irmgard. Country Report from Germany. In: Nippert, Irmgard., Clausen, Hans Frets, Petra, Niermeijer Martinus and Modell Michael (eds.), *Evaluating Cystic Fibrosis Carrier Screening Development in Northern Europe: Denmark, The Federal Republic of Germany, the Netherlands, and the United Kingdom.* Münster: Westf. Wilhelms-Universität, Women's Health Research Series 1998, pp. 43–70.

Nippert, Irmgard and Wertz, Dorothy. Grundprinzipien bei der Anwendung genetischer Testverfahren. In: Winter, S./Fenger, H./Schreiber, H.-L (eds.), *Genmedizin und Recht.* München: Verlag C.H. Beck, 2001.

Nippert, Irmgard and Wolff, Gerhard. Consensus and Variation among Medical Geneticists and Patients on the Provision of the New Genetics in Germany. Data from the 1994–6 Survey among Medical Geneticists and Patients. In: Nippert, I, Neitzel, H and Woldd, G. *The New Genetics: From Research into Health Care: Social and Ethical Implications for Users and Providers.* Berlin: Springer, 1999, pp. 95–106.

Nippert, Irmgard and Wolff, Gerhard. Germany. In: Wertz Dorothy and Fletcher John (eds), *Genetics and Ethics in Global Perspective.* Dordrecht: Kluwer academic publishers, 2004, pp. 223–235.

Nir, Uri. "Family tree of diseases". *Haaretz*, 11.05.2001 (in Hebrew).

Novas, Carlos and Rose, Nikolas. Genetic risk and the birth of the somatic individual. *Economy and Society.* 2000; 29(4): 485–513.

Oakley, Ann. *The Captured Womb: A History of the Medical Care of Pregnant Women.* Oxford: Basil Blackwell, 1984.

Ophir, Adi. Moral Technologies: Managing Disaster and Forsaking Life. *Theory and Criticism.* 2003a; 22: 67–103 (in Hebrew).

Ophir Adi. Between Sacred and Abandoned Life: Instead of a Foreword to Homo Sacer. In: Shai Lavi (ed.), *Technologies of Justice: Law, Science and Society.* Tel-Aviv: Ramot publishing house, university of Tel-Aviv, 2003b, pp. 353–395 (in Hebrew).

Parens Erik and Asch Adrienne. *Prenatal Testing and Disability Rights.* Washington DC, Georgetown University Press 2000.

Parsons, Talcot. The professions and social structure. In: Talcot Parsons (ed.), *Essays in Sociological Theory.* Glencoe, IL, Free Press, 1954, pp. 34–49.

Paul, Diane.B. Eugenic Anxieties, Social Realities and Political Choices. *Social Research.* 1992; 59: 663–683.

Paul, Diane B. *Controlling Human Heredity: 1865 to the Present.* New Jersey: Humanities Press, 1995.

Paul, Diane B. *The Politics of Heredity: Essays on Eugenics, Biomedicine, and the Nature-Nurture Debate.* New York: State University of N.Y. Press, 1998.

Payer, Lynn. *Medicine and Culture: Varieties of Treatment in the US, England, West Germany and France.* New York: Henry Holt, 1988.

Penal Law, 1977. Interruption of Pregnancy 312–321. (For an English translation see Falk, W. Zeev. The New Abortion Law of Israel. *Israel Law Review.* 1978; 13: 109–110.

Peres, Yohanan and Katz, Ruth. The Family in Israel: Change and continuity. In: Lea Shamgar-Handelman and Rivka Bar-Yosef (eds.), *Families in Israel.* Academon – The Hebrew University Student's Printing and Publishing House in Jerusalem, 1991, pp. 9–32 (in Hebrew).

Petersen, Kerry. Criminal Abortion Laws: An impediment to Reproductive Health? In: Ian Freckelton and Kerry Petersen (eds.), *Controversies in Health Law.* Sydney: The Federation Press, 1999.

Petersen, Kerry. Abortion: Medicalisation and Legal Gatekeeping. *Journal of Law and Medicine.* 2000; 7: 267.

Pinch T.J., Bijker W.E. The Social Construction of Facts and Artifacts: Or how the Sociology of Science and the Sociology of Technology might benefit each other. In: Bijker, W.E., Hughes, T.P., Pinch, T.J. (eds.), *The Social Construction of Technological Systems: New Direction in the Sociology and History of Technology.* Cambridge MA: MIT Press, 1987.

Portugese, Jacqueline. *Fertility Policy in Israel. The Politics of Religion, Gender and Nation.* London: Praeger, 1998.

Positionspapier der Gesellschaft fuer Humangenetik e.V., *Med. Genetik* 1/1996, pp. 125–131.

Prainsack, Barbara. The Politics of Life: 'Negotiating Life' in Israel. Embryonic Stem Cells, and Human Cloning. PhD thesis, University of Vienna, Austria, 2004.

Prainsack, Barbara. 'Negotiating Life': The Regulation of Human Cloning and Embryonic Stem Cell Research in Israel. *Social Studies of Science*. 2006a; 36(2): 173–205.

Prainsack, Barbara. The 'conflict of conflicts': Human Reproductive Cloning and the Creation of New Citizens. Paper presented at the ECPR Joint Session of Workshops, 2006b.

Prainsack Barbara and Firestine Ofer. Science for survival: Biotechnology regulation in Israel. *Science and Public Policy*. 2006; 33(14): 33–46.

Prainsack Barbara and Siegal Gil. The Rise of Genetic Couplehood? A Comparative View of Premarital Genetic Testing. *BioSocieties*. 2006; 1: 17–36.

Prainsack Barbara and Tim D. Spector. Twins: A cloning experience. *Social Science and Medicine*. 2006: 63/10: 2739–2752.

Pregnancy Terminations According to the Law: Trends in the Activities of Abortion Comities 1990–1999. Jerusalem: The state of Israel, Health Ministry, September, 2000 (in Hebrew).

Pregnant Women's and Families' Aid Amendment Act (Schwangeren- und Familienhilfeänderungsgesetz, 1995, BGB1 I 1050-57)

Pritchard, Megan. Can there be such a thing as a 'wrongful birth'? *Disability and Society* 2005; 20(1): 81–93.

Proposed International Guidelines on Ethical Issues in Medical Genetics and Genetic Services. 1998. Human Genetics Program, World Health Organization.

Public Survey Concerning Science in Israel, 2002. Handed to Prof. Hagit Messer-Yaron, The Israeli science, culture and sports ministry. (Can be received upon personal request from Prof. Messer-Yaron).

Rabino, Isaac. German Genetic Engineering Scientists and the German Public: Complementary Perceptions in a Changing European Context. *Public Understanding of Science*. 1994; 3(4): 365–384.

Rabinow, Paul. Artificiality and enlightment: From sociobiology to biosociology. In: Crary, J. and Kwinter, S. (eds.), *Zone 6: Incorporations*. Cambridge, MA: MIT Press, 1992.

Rabinow, Paul. *Essays on the Anthropology of Reason*. Princeton, New Jersey: Princeton University Press, 1996.

Rabinow, Paul. *French DNA: Trouble in Purgatory*. Chicago: The University of Chicago Press, 1999.

Ram, Uri. Between Nation and Corporations: Liberal Post-Zionism in the Global age. *Israeli Sociology*. 1999; 2(1): 99–145 (in Hebrew).

Rapp Rayna. *Testing Women, Testing the Fetus: The Social Impact of Amniocentesis in America*. New York: Routledge, 1999.

Rat der Evangelischen Kirche Deutschlands (EKD), 'Der Schutz menschlicher Embryonen darf nicht eingeschränkt werden', 22 May 2001.http://www.ekd.de/presse/39/

Rau, Johannes. Will Every-thing turn out well? For progress befitting humanity. Berlin Address by Federal President Johannes Rau in the Otto-Braun-Saal of the Berlin State Library, 18 May, 2001.

Raz, Aviad, E. 'Aysha': Genetics education and community engagement in a consanguineous Arab-Bedouin population in Israel. *Health: An interdisciplinary Journal for the Social Study of Health, Illness and Medicine*. 2003; 7(4): 439–461.

Raz, Aviad. 'Important to Test, Important to Support': Attitudes toward Disability Rights and Prenatal Diagnosis among Leaders of Support Groups for Genetic Disorders in Israel. *Social Science and Medicine*. 2004; 59: 1857–1866.

Reed Sheldon, C. "A Short History of Genetic Counseling". *Social Biology*. 1974; 21:332–339.

Remennick, Larissa. Childless in the land of Imperative Motherhood: Stigma and Coping among Infertile Israeli Women. *Sex Roles*. 2000; 43(11/12): 821–841.

Remennick, Larissa. The Quest after the Perfect Baby: Why Do Israeli Women Seek Prenatal Genetic Testing? *Sociology of Health and Illness*. 2006; 28(1): 21–53.

Report of the Public-Professional Committee to Examine In Vitro Fertilization. 1994. Jerusalem: Israel Ministry of Justice (in Hebrew)

Resnik, Julia and Frenkel, Michal. From Critical Sociology to the Sociology of Criticism: Comments on the Pragmatistic Sociology of Luc Boltanski. *Theory and Criticism*. 2000; 17(Fall):101–122 (in Hebrew).

Richardt, Nicole. A Comparative Analysis of Embryological Research Debate in Great Britain and Germany. *Social Politics*. 2003 Spring: 86—128.

Richtlinien zur pränatalen Diagnostik von Krankheiten und Krankheitsdispositionen, Bekanntmachungen der Bundesärztekammer, *Dt. Ärztebl.* 1998; 95: A-3236-3242 [Heft 50]

Robertson, Roland. *Globalization: Social Theory and Global Culture*. London: Sage, 1993.

Robertson, Roland. Glocalization: Time-Space and Homogeneity-Heterogeneity. In: Mike Featherstone, Scott Lash and R. Robertson (eds.), *Global Modernities*. London: Sage, 1995, pp. 25–44.

Rose Nikolas. The Politics of Life itself. *Theory Culture and Society*. 2004; 18(6): pp. 1–30.

Sarat, Austin and Felstiner, William.L.F. *Divorce lawyers and their Clients: Power and Meaning in the Legal Process*. New York: Oxford University Press, 1995.

Sawicki, Jana. Disciplining Mothers: Feminism and the New Reproductive Technologies. In: Janet Price and Margrit Shildrick (eds.), *Feminist Theory and the Body: A Reader*. Edinburgh: Edinburgh University Press, 1999, pp. 190–202.

Schiff, Gary, S. The Politics of Fertility Policy in Israel. In: Ritterband, P. (ed.), *Modern Jewish Fertility*. Leiden: Brill, 1981, pp. 255–278.

Schlegel, Christina. P. Landmark in German Abortion Law: The German 1995 Compromise Compared with English Law. *International Journal of Law, Policy and the Family*. 1997; 11: 36–62.

Schoen, Donald. A and Rein, Martin. *Frame Reflection: Toward the Resolution of Intractable Policy Controversies*. Basic Books, NY 1994.

Schroeder-Kurth T.M. Individual genetic counseling or eugenics? In: Obe, G (ed.), *Advances in Mutagenesis Research*. Berlin: Springer-Verlag, 1990, pp. 231–247.

Schroeder-Kurth, T.M. and J Huebener. Ethics and Medical Genetics in the Federal Republic of Germany. In: Wertz, D.C. and Fletcher, J.C. (eds.), *Ethics and Human Genetics: A cross Cultural Perspective*. Berlin: Springer-Verlag, 1989, pp. 156–175.

Schwangerschaftskonfliktgesetz, SchKG, paragraph 12 . 1995

Scranton, Philip. Determinism and Indeterminacy in the History of Technology. In: Leo Marx and Merritt R. Smith (eds.), *Does Technology Drive History? The Dilemma of Technological Determinism*. Cambridge Massachusetts: The MIT Press, 1994, pp. 143–168.

Sered Susan, *What Makes Women Sick? Maternity, Modesty, and Militarism in Israeli Society*. Brandeis Series on Jewish Women. Boston: Brandeis University Press, 2000.

Shadmi, Haim. "A New Report: The Rate of Genetic Birth Defects in Israel is High". *Haaretz*, 08.05.2001a (in Hebrew).

Shadmi, Haim. "New-Born's Death Rates are Twice as High in Hadera District among the Arabs". *Haaretz*, 14.05.2001b (in Hebrew).

Shalev, Carmel. Fertility law and individual's right to become parents. In: F. Raday, C. Shalev, M. Libman-Kooly. Schocken (eds.), *Women's Status in Israeli Law and Society*. Publishing House Ltd. Tel Aviv, 1995, pp. 508–533.

Shapira, Amos. 'Wrongful Life' Suits by Defective Newborns for Faulty Genetic Counseling. *The Human Genome Project: Legal, Social and Ethical Implications. Proceedings of an International Workshop. Jerusalem, 5 July 1995*. Jerusalem 1997: The Israel Academy of Sciences and Humanities.

Shenhav, Yehuda. *The Organization Machine: A Critical Inquiry into the Foundations of Management Theory*. Jerusalem: Schoken, 1995 (in Hebrew).

Shenhav, Yehuda. *Manufacturing Rationality: The Engineering Foundations of the Managerial Revolution*. Oxford: Oxford University. Press, 1999.

Shenhav, Yehuda. *The Arab-Jews: Nationalism, Religion and Ethnicity*. Tel-Aviv: Am Oved, 2003 (in Hebrew).

Shenhav, Yehouda.. Nationalism was never Modern – or Secular: On Hybridization and Purification in Latour. *Theory and Criticism*. 2005; 26: 75–89.

Sher, Carron et al. Factors Affecting Performance of Prenatal Genetic Testing by Israeli Jewish Women. *American Journal of Medical Genetics*. 2003 120A: 418–422.

Shildrick, Margrit and Price, Janet. Splitting the Difference: Adventures in the Anatomy and Embodiment of Women. In: Griffin, Gabriele. (eds.), *Stirring It: Challenges for Feminism*. Taylor & Francis 1994, Chapter 9.

Shilling, Chris. *The Body and Social Theory*. London: Sage Publications, 1993.

Shilling, Chris and Mellor, Philip, A. The Sociological Ambition: Elementary Forms of Social and Moral Life. London: Sage, 2001.

Smith, Merritt Roe and Marx, Leo. *Does Technology Drive History? The Dilemma of Technological Determinism*. Cambridge Massachusetts: The MIT Press, 1994.

Spakes, Patricia. Women, Work and Babies: Family-Labor Market Policies in Three European countries. *Affilia*. 1995; 10(4 Winter): 369–397.

Sperling, K, Pelz, J., Wegner R-D, Dörries, A., Grüters A., Mikkelsen M. Significant increase in trisomy 21 in Berlin nine months after the Chernobyl reactor accident. Temporal correlation or causal reaction?. *BMJ*. 1994; 309: 158–162.

Staudenmair, J.M. *Technology's Storytellers: Reweaving the Human Fabric*. Cambridge MA: The MIT Press, 1985.

Steinberg, Abraham. Fertility and Infertility. *Medical Halakhic Encyclopedia*. Jerusalem: The Schlezinger institute for the study of medicine according to the bible, 1988–96, pp. 372.

Steinberg, Abraham. 'Abortion', *Medical Halakhic Encyclopedia*. Jerusalem: The Schlezinger Institute for the Study of Medicine according to the Bible, 1991, 2: 90–91.

Steinberg, Avraham. The Meaning of Suffering: A Jewish Perspective. In: Edmund D. Pellegrino and Alan I. Faden (eds.), *Jewish and Catholic Bioethics: An Ecumenical Dialogue*, Washington, DC 1999, p. 80f.

Stoler-Liss, Sachlav. *Zionist Baby and Child Care: Anthropological Analysis of Parents' Manuals*. Tel Aviv University: M.A. Thesis, 1998 (in Hebrew).

Stoler-Liss, Sachlav. Mothers Birth the Nation: The Social Construction of Zionist Motherhood in Wartime Israeli Parents' Manuals. *Nashim*. 2003; 6 (Fall): 104–118.

Strasberg-Cohen, Tirza. The Status of the Embryo under Israeli law. In: Blazer Shraga and Zimmer, Etan Z (eds.), *The Embryo: Scientific Discovery and Medical Ethics*. Basel: Karger, 2005.

Study Commission on "Law and Ethics in Modern Medicine", Final Report, German Bundestag 14th Legislative Period. Document 14/9020, 14 May, 2002. http://www.bundestag.de/parlament/kommissionen/archiv15/ethik_med/archiv/schlussbericht_14_WP_engl.pdf.

Sturman, Randy Linda. *Six Lives in Jerusalem: End-of-Life decisions in Jerusalem – Cultural, Medical, Ethical and Legal Considerations*. Dordrecht: Kluwer, 2003.

Sutton, Agneta. The British Law on Assisted Reproduction: A Liberal Law by Comparison with Many Other European Laws. *Ethics and Medicine*. 1996; 12(2): 41–45.

Swidler, Ann. Culture in Action: Symbols and Strategies. *American Sociological Review*. 1986; 51: 273–286.

Teman, Elly. The Medicalization of 'Nature' in the 'Artificial Body': Surrogate Motherhood in Israel. *Medical Anthropology Quarterly*. 2003; 17, 1: 78–98.

The Science and Technology Committee. PGD for sex selection and other medical purposes, Israeli parliament ('Knesset'), protocol number 124, 2005.

Thomas, Mark G., Parfitt Tudor, Weiss Deborah A, Skorecki Karl, Wilson James F, le Roux Magdel, Bradman Neil and Goldstein David B. Y Chromosomes Traveling South: The Cohen Modal Haplotype and the Origins of the Lemba—the 'Black Jews of Southern Africa'. *American Journal of Human Geneics*. 2000; 66: 674–686.

Timmermans Stefan and Berg Marc, "The practice of medical technology", *Sociology of Health and Ilness*. 2003; 25: 97–114.

Tomlinson John, *Globalization and Culture*. Chicago, 1999.

Toulmin, Stephen. *Cosmopolos: The Hidden Agenda of Modernity*. New York: Free Press, 1990.

Trabelsi-Hadad, Tamar. "Today's Youth". Yedioth Ahronoth, 26.12.2005. ("Hanoar Shel Hayom", in Hebrew). p. 13.

Traubman, Tamara. "Ben-Gurions' Genome project" *Haaretz*, 6.5.2003a (in Hebrew).

Traubman, Tamara. "Experts warn about unnecessary abortions, but the hospitals continue to recommend the test". *Haaretz*, 10.6.2003b (in Hebrew).

Usborne, Cornelie. *The Politics of the Body in Weimar Germany: Women's Reproductive Rights and Duties*. Ann Arbor, the University of Michigan Press, 1992.

Van Zuuren, F.J. The standard of neutrality during genetic counseling: An empirical investigation. *Patient Education and Counseling*. 1997; 32: 69–79.

Vardimon, David and Ben-Rafael, Zion. *A Guide to Prenatal Counseling*. Tel-Aviv: Schoken Publishing House, 1999 (in Hebrew).

Von Cube, Alex. *The Fertility Decline in the Federal Republic of Germany: A Political Analysis*. Ph.D. dissertation submitted to the Faculty of Graduate School of the University of Maryland, USA, 1989.

Wahrman, Miryam Z. *Brave New Judaism: When Science and Scripture Collide*. Hanover, MA: Brandeis University Press, 2002.

Walter, J. James. Theological Issues in Genetics. *Theological Studies*. 1999; 60(1): 124–134.

Ward, Linda, M. Whose right to choose? The 'New' genetics, prenatal testing and people with learning difficulties. *Critical Public Health*. 2002; 12(2): 187–200.

Weber, Max. *Economy and Society*. Berkeley: University of California Press, 1968.

Weiss, Meira. *Conditional Love: Parents' Attitudes towards Handicapped Children*. Westport, Conn: Bergin and Garvey, 1994.

Weiss, Meira. *The Chosen Body: The Politics of the Body in Israeli Society*. Stanford: Stanford University Press, 2002.

Wegner, R.D. and Becker, R. Prenatal Diagnosis in Germany. *European Journal of Human Genetics*. 1997; 5(1): 32–38.

Welsh, Helga A., Pickel, Andreas. and Rosenberg, Dorothy. East and West German Identities: United and Divided? In: Konrad H. Jarausch (ed.), *After Unity: Reconfiguring German Identities*. Oxford: Berghahn Books, 1997, pp. 103–136.

Wertz, Dorothy C. Is there a 'Women's Ethic' in Genetics: A 37-Nation Survey of Providers. *JAMWA*. 1997a; 52(1): 33–38.

Wertz, Dorothy, C. Society and the not so-new genetics: 'What are we afraid of?' Some future predictions from a social scientist. *The Journal of Contemporary Health Law and Policy*. 1997b; 13: 299–346.

Wertz, Dorothy C. Eugenics is Alive and Well. In: R. Falk and D.B. Paul (eds.), *Science in Context*. 1998, Autumn-Winter; 11(3–4):493–510.

Wertz, Dorothy, C. 1999. "Ethics and genetics in International Perspective: Results of a Survey ", in I. Nippert , H. Neitzel and G. Woldd (eds), *The New Genetics: From Research into Health Care*. Berlin: Springer. pp. 75–94.

Wertz, Dorothy C. and Fletcher, John C. Genetics Approach Ethics: A Survey in 37 Nations. *Social Science, Ethics & Law*. Shriver Center, 200 Trapelo Rd, Waltham MA 02254, USA, 1993–5.

Wertz, Dorothy C. and Fletcher, John C. *Genetics and Ethics in Global Perspective*. Dordrecht–Boston–London: Kluwer Academic Publishers, 2004.

Wildung Harrison, Beverly. Abortion, Religious Tradition: Protestant Perspectives. In Warren T. Reich (ed.), *Bioethics: Sex, Genetics and Human Reproduction*. Macmillan compendium: Single volume encyclopedia, New York, 1998.

Wilson J.M.G., Junger G. *Principles and Practice of Screening for Disease*. Geneva: World Health Organization, 1968.

Wolff, Gerhard. and Jung, Christine. Nondirectiveness and Genetic Counseling. *Journal of Genetic Counseling*. 1995; 4(1): 3–25.

Wright, Peter and Treacher, Andrew. (eds.), *The Problem of Medical Knowledge: Examining the Social Construction of Medicine*. Edinburgh University Press, 1982.

Wuerth, Andrea. Re-Unification and Reproductive Rights: Abortion in the German Public Sphere, 1989–1990. *Working paper, Center for European Studies*. University of North Carolina, Chapel Hill, 1997.

Wynn, Francine. Nursing and the concept of life: Towards an ethics of testimony. *Nursing Philosophy*. 2002; 3: 120–132.

Yasur-Beit-Or, Meital, "For the first time, a committee allowed a couple to choose the sex of their child". http://www.ynet.co.il/articles/1,7340,L-3220136,00.html. Retrieved 24.02.2006

Yishai, Yael. Public Ideas and Public Policy: Abortion Policies in Four Democracies. *Comparative Politics*. 1993; 25: 207–228.

Yonay, Yuval and Spivak, Dori. Between Silence and Damnation: The Construction of Gay Identity in the Israeli Legal Discourse. *Israeli Sociology*. 1999; 1(2): 257–295.

Yuval-Davis, Nira. The bearers of the collective: Women and religious legislation in Israel. *Feminist Review*. 1980; 4: 15–27.

Yuval-Davis, Nira. The Jewish Collectivity. In: Salman, M. (ed.), *Women in the Middle East*. London: Zed Books, 1987.

Yuval-Davis, Nira. *Gender and Nation*. London: Sage Publications, 1997.

Zaitsov v. Katz, C.A 540/82, 40 P.D. (2) 85.

Zelizer, Viviana A. *Pricing the Priceless Child: The Changing Social Value of Children*. Basic Books, 1985 NY.

Zilber, Carmel. "Who really needs genetic tests", *Yediot Aharonot, zmanim modernim*. 06.01.1999 (in Hebrew).

Zimmermann Mirjam, Zimmermann Ruben. Gibt es ein Recht auf ein gesundes Kind? *Deutsches Aerzteblatt*. 2000; Dezember, Heft 51–52.

Zimran, A., Zaizov, R., Zlotogora, J. Large scale screening for Gaucher's disease in Israel – a position paper by the National Gaucher Committee of the Ministry of Health. *Harefuah*. 1997; 133: 107–108 (in Hebrew).

Zlotogora, Joel and Chemke, Juan. Medical Genetics in Israel. *European Journal of Human Genetics*. 1995; 3: 147–154.

Zlotogora, Joël and Leventhal, Alex. Screening for Genetic Disorders among Jews: How should the Tay-Sachs Screening program be Continued? *Israeli Medical Association Journal*. 2000; 2: 665–667.

Zohar, Noam J. From Lineage to Sexual Mores: Examining 'Jewish Eugenics'. *Science in Context*. 1998; 11(3–4): 575–585.

INDEX

International Library of Ethics, Law, and the New Medicine

1. L. Nordenfelt: *Action, Ability and Health.* Essays in the Philosophy of Action and Welfare. 2000 ISBN 0–7923–6206–3
2. J. Bergsma and D.C. Thomasma: *Autonomy and Clinical Medicine.* Renewing the Health Professional Relation with the Patient. 2000 ISBN 0–7923–6207–1
3. S. Rinken: *The AIDS Crisis and the Modern Self.* Biographical Self-Construction in the Awareness of Finitude. 2000 ISBN 0–7923–6371–X
4. M. Verweij: *Preventive Medicine Between Obligation and Aspiration.* 2000 ISBN 0–7923–6691–3
5. F. Svenaeus. *The Hermeneutics of Medicine and the Phenomenology of Health.* Steps Towards a Philosophy of Medical Practice. 2001 ISBN 0–7923–6757–X
6. D.M. Vukadinovich and S.L. Krinsky: *Ethics and Law in Modern Medicine.* Hypothetical Case Studies. 2001 ISBN 1–4020–0088–X
7. D.C. Thomasma, D.N. Weisstub and C. Hervé (eds.): *Personhood and Health Care.* 2001 ISBN 1–4020–0098–7
8. H. ten Have and B. Gordijn (eds.): *Bioethics in a European Perspective.* 2001 ISBN 1–4020–0126–6
9. P.-A. Tengland: *Mental Health.* A Philosophical Analysis. 2001 ISBN 1–4020–0179–7
10. D.N. Weisstub, D.C. Thomasma, S. Gauthier and G.F. Tomossy (eds.): *Aging: Culture, Health, and Social Change.* 2001 ISBN 1–4020–0180–0
11. D.N. Weisstub, D.C. Thomasma, S. Gauthier and G.F. Tomossy (eds.) *Aging: Caring for our Elders.* 2001 ISBN 1–4020–0181–9
12. D.N. Weisstub, D.C. Thomasma, S. Gauthier and G.F. Tomossy (eds.): *Aging: Decisions at the End of Life.* 2001 ISBN 1–4020–0182–7
    (Set ISBN for vols. 10–12: 1–4020–0183–5)
13. M.J. Commers: *Determinants of Health: Theory, Understanding, Portrayal, Policy.* 2002 ISBN 1–4020–0809–0
14. I.N. Olver: *Is Death Ever Preferable to Life?* 2002 ISBN 1–4020–1029–X
15. C. Kopp: *The New Era of AIDS.* HIV and Medicine in Times of Transition. 2003 ISBN 1–4020–1048–6
16. R.L. Sturman: *Six Lives in Jerusalem.* End-of-Life Decisions in Jerusalem-Cultural, Medical, Ethical and Legal Considerations. 2003 ISBN 1–4020–1725–1
17. D.C. Wertz and J.C. Fletcher: *Genetics and Ethics in Global Perspective.* 2004 ISBN 1–4020–1768–5
18. J.B.R. Gaie: *The Ethics of Medical Involvement in Capital Punishment.* A Philosophical Discussion. 2004 ISBN 1–4020–1764–2
19. M. Boylan (ed.): *Public Health Policy and Ethics.* 2004 ISBN 1–4020–1762–6; Pb 1–4020–1763–4
20. R. Cohen-Almagor: *Euthanasia in the Netherlands.* The Policy and Practice of Mercy Killing. 2004 ISBN 1–4020–2250–6
21. D.C. Thomasma and D.N. Weisstub (eds.): *The Variables of Moral Capacity.* 2004 ISBN 1–4020–2551–3
22. D.R. Waring: *Medical Benefit and the Human Lottery.* An Egalitarian Approach. 2004 ISBN 1–4020–2970–5

International Library of Ethics, Law, and the New Medicine

23. P. McCullagh: *Conscious in a Vegetative State? A Critique of the PVS Concept.* 2004 ISBN 1–4020–2629–3
24. L. Romanucci-Ross and L.R. Tancredi: *When Law and Medicine Meet: A Cultural View.* 2004 ISBN 1–4020–2756–7
25. G.P. Smith II: *The Christian Religion and Biotechnology.* A Search for Principled Decision-making. 2005 ISBN 1–4020–3146–7
26. C. Viafora (ed.): *Clinical Bioethics.* A Search for the Foundations. 2005 ISBN 1–4020–3592–6
27. B. Bennett and G.F. Tomossy: *Globalization and Health.* Challenges for health law and bioethics. 2005 ISBN 1–4020–4195–0
28. C. Rehmann-Sutter, M. Düwell and D. Mieth (eds.): *Bioethics in Cultural Contexts.* Reflections on Methods and Finitude. 2006 ISBN1–4020–4240–X
29. S.E. Sytsma, Ph.D.: *Ethics and Intersex.* 2006 ISBN 1–4020–4313–9
30. M. Betta (ed.): *The Moral, Social, and Commercial Imperatives of Genetic Testing and Screening.* The Australian Case. 2006 ISBN 1–4020–4618–9
31. D. Atighetchi: *Islamic Bioethics: Problems and Perspectives.* 2006 ISBN 1–4020–4961–7
32. V. Rispler-Chaim: *Disability in Islamic Law.* 2006 ISBN 1–4020–5052–6
33. Y. Denier: *Efficiency, Justice and Care.* Philosophical Reflections on Scarcity in Health Care. 2007 ISBN 978–1–4020–5213–2
34. Y. Hashiloni-Dolev: *A Life (Un)Worthy of Living.* Reproductive Genetics in Israel and Germany. 2007 ISBN 978–1–4020–5217–0

Printed in the United States
118467LV00003B/204/A